D0778824

APR 1 2 1995

5/7-99
5/11-99

Engineering

for PROFIT

SUCCESSFUL MARKETING OF HI-TECH PRODUCTS AND SYSTEMS

Engineering
for PROFIT

SUCCESSFUL MARKETING
OF HI-TECH PRODUCTS
AND SYSTEMS

O. Judson Strock

CRC Press
Boca Raton   Ann Arbor   London   Tokyo
A SOLOMON PRESS BOOK

OC 0495 J.c.

Library of Congress Cataloging-in-Publication Data

Strock, O. J. (O. Jud), 1923-
  Engineering for profit : successful marketing of high-tech
  products and systems / O. Judson Strock.
      p.   cm.
  "A Solomon Press book."
  Includes index.
  ISBN 0-8493-8974-7
  1. High technology--Marketing.   2. High technology industries--
    Marketing.   3. Engineering services marketing.   4. Service
    industries--Marketing.   I. Title.
  HC79.H53S77 1994
  620'.0068'—dc20
  DNLM/DLC
  for Library of Congress   94–16831
                                                                  CIP

Contents

Preface

THE MATERIAL IN THIS BOOK represents an accumulation of observations from my thirty years of marketing experience in various job assignments, during which time much was learned from the supervisors, subordinates, sales engineers, marketing managers, cost accountants, word processor operators, proposal managers, customers, competitors, and others in the marketing arena. All those contributions are appreciated, either because of their positive content or because they tended to be "WACKY" examples of what not to do. I am also thankful for the feedback I received at seminars where I presented the ideas developed in this book.

I am grateful for the contributions of Dr. John Wood of Loral Data Systems, who worked closely with the publisher in formatting the book. John Talbot added several appropriate observations based on his proposal experience and wrote the section on advertising to incorporate some of his experiences in his own advertising agency.

Sandy Basham of Electronic Publishing at Loral Data Systems was most helpful in preparing camera-ready publication material. Jon Wolf and his Graphics Group provided most of the illustrations.

This is my second book under the skilled publishing management of Sidney Solomon and Raymond Solomon. I am indeed grateful for the advice and encouragement of Joel Claypool of CRC Press.

This book uses many examples, most of them electronics-related. This is done because it provides consistency, because the electronics business is the most widely based high-technology industry, and because my experience has been predominantly in electronics. The reader, however, can extrapolate very easily to other high-technology fields; the methods and examples are equally appropriate for those fields. Some of the types of proposals to which the book relates are:

Airplanes
Automatic Test Equipment
Development Programs
Digital Computers
   Hardware/Peripherals
   Software
Electronic Instruments
Electronic Warfare Equipment
Engines/Turbines
Fiber Optics Systems
Helicopters
Laser Equipment
Magnetic Recorders
Medical Instruments
   and Equipment
Missiles and Rockets

Motor Vehicles
Navigational Systems
Photographic Equipment
Radar Systems
Radio Systems
Refrigeration/Air Conditioning
Research Programs
Satellites and Space Vehicles
Architectural/Engineering
Ships/Marine Equipment
Surveillance Equipment
Telemetry Systems
Telephone Systems
Television Equipment
   and Systems

I hope that each reader will find here the information which is needed for more efficient, more productive proposal preparation, product planning, and marketing in general.

O. Judson Strock

1

Finding and Understanding Your Customer

ENGINEERING FOR PROFIT is addressed to several types of people in any high-technology company. For the engineer, programmer, or other highly skilled professional, it is a labor-saving aid, making him or her more effective in the support of marketing activities for the benefit of everyone in the company. For the proposal manager, it provides many step-by-step techniques to make the proposal team more effective. For support personnel (manufacturing, procurement, documentation, and so on), it illustrates their contribution to making proposals top-quality sales documents. For marketing managers, it is the technical manual they would write if they could find the time. For the company president or division manager, it is the tool for selling more products and systems with bid and proposal costs that are the same or even lower.

The job of marketing high-technology products or systems often calls for the collaboration of a person who is trained and experienced as an engineer or in another professional capacity, but has been drafted part time for marketing support. He or she may be involved in design for 90% of the time and in marketing support for the other 10%, or may have transferred into a mode where marketing is more dominant. Why is it necessary for some highly skilled persons to leave their primary skills behind, even for 10% of their time, and function in this alien territory called "marketing"? The answer is very simple: high-technology products and systems often have to be marketed with the help of high-technology people. High-technology companies cannot depend totally on nontechnical personnel for their marketing and sales efforts. Some of the best engineering talent must frequently be drafted in the quest for better products, better proposals, and more business.

For that experienced professional who is nevertheless a newcomer to marketing, this book gives a helping hand; it:

1. Shows how to prepare for a technical proposal, defines points for emphasis, and lays the groundwork needed to reach a particular customer or agency.

2. Discusses the organization of the proposal team and how best to motivate them.

3. Talks about hardware and software sections and how to achieve that engineer-to-engineer communication in proposals.

4. Elaborates on some guidelines for estimating costs, to help land contracts and make money.

5. Emphasizes the ways to get the most proposal for the time and money spent.

6. Uses dozens of practical examples to illustrate points.

7. Defines the customer, whose existence is the life blood of the company.

8. Lays down useful guidelines for product planning, sales literature, and applications notes.

9. Outlines some approaches to advertising, new product releases, trade shows, and other ways to locate and attract new customers.

10. Gives an introduction to technical papers—why write them, how to prepare them, and some rules of presentation.

11. Gets down to details with case histories—what to do and what not to do.

12. Provides exercises to encourage and enhance individual and group study.

ORGANIZATION

This book is organized to show the various aspects of high-technology marketing.

Chapter 1. Finding and Understanding Your Customer. This chapter explains the purpose of the book and its benefits to various categories of personnel, helping each reader to use the material for his or her good and the good of the company. It then summarizes the individual chapters, showing how each fits into an overall pattern of value to the technical professional who has been enlisted for marketing support.

Because a "customer" must be understood and appreciated in order to be won, the chapter also gives an overview of the corporate customer and an introduction to the people who constitute a customer group.

Chapter 2. Bid Decisions and Marketing Strategy. The importance of correct bid decisions to the success of any company is outlined. Doors will close if this aspect of business life is not handled properly. The technique of bid decision analysis is explained in detail, with a discussion of the many factors that must be integrated into the analysis.

When a positive decision is reached on any bid opportunity, a powerful "win strategy" must be developed to supplement the decision and guide the proposal manager and team members to winning the contract. Since the proposal budget is an important factor in company operation, developing and monitoring the budget is discussed in detail.

Chapter 3. Getting Started. The most important person in the bid sequence is the proposal manager. His or her selection and support are outlined here, as are the basics of selecting and motivating the proposal team.

In preparing for a meaningful kickoff meeting, the proposal manager will work with the win strategy to define the significant differences between the company and its competitors and to develop a means for amplifying the positives and overcoming the negatives. A storyboard sequence is developed to complement the proposal outline, so that the writers can move most efficiently toward their common goal.

The kickoff meeting, which makes up the most significant few hours in the whole bid sequence, is described in detail to show how a team can be directed and motivated to success.

Chapter 4. Your Proposal—The Technical Parts. The reader is shown how to use the win strategy, proposal outline, storyboards, and other material from the kickoff meeting to write the technical parts of the proposal. Resources from previous proposals and other company files are an important part of this process.

A concise, meaningful abstract is developed, followed by the introductory section of the proposal. Then an overview of the system or device being offered will be written, as outlined here. Technical details are developed to define the pieces of the system, and a specification analysis is prepared to show paragraph-by-paragraph conformance to the specification.

Chapter 5. Your Proposal—Those Other Parts. Nontechnical sections, which are often down-played in importance by the technical professional, are defined here in a correct perspective. First, the management of the proposed contract must be planned in detail and described for the customer. Resumes must be prepared for all the key personnel on the program. The various support functions defined in the program management section (such as quality assurance, test, documentation, and service) must be described for the customer. Corporate capabilities, such as the buildings, laboratories, instruments, and data processing systems, must be described.

Details that do not fit into the flow of the proposal but are important in establishing capabilities should be shown in a well-planned, well-organized appendix to the proposal.

Chapter 6. Cost Estimating and Pricing Strategy. Estimating costs and determining price are discussed in detail, with description of a cost-estimating matrix and its use in the determination of probable costs on the job. The various elements that enter into determining a selling price are outlined to provide a minimum risk of being disqualified due to an excessive bid for equipment and services.

Chapter 7. Putting It All Together. After all the details have been handled, this is the ribbon to tie it together. General wrap-up information is considered first, and then comes the important, but often hastily constructed, cover letter. The chapter also includes some time- and money-saving techniques that are valuable to the proposal team during these hectic weeks of activity.

Chapter 8. After the Proposal Is Mailed—The Follow-up. The actions that should take place as soon as a proposal is mailed, while the team is together and the subject is on everyone's mind, are described here.

For times when there is an opportunity for formal verbal presentations to the customer, preparation for that very important day is outlined.

Finally, when the customer's decision on the procurement is announced, there should be a self-analysis of the team's winning or losing performance, with specific ideas for improvements that should be used on the next proposal. A critique method is outlined.

Chapter 9. Case History of a Winning Proposal. This is a slightly abridged case history of an actual winning proposal sequence, with examples of most of the documents written. It starts with the advance work of the sales engineer and in-plant technical professionals and moves on to the bid decision and the design of a marketing strategy.

The proposal kickoff meeting is described, with the proposal manager's handout shown. Parts of the technical sections of the proposal are shown, along with excerpts from the nontechnical section. The costing matrix is shown, as is a discussion of the pricing decision. A sample cover letter is shown and the formal presentation is described, as is the post-award analysis for this winning proposal.

Chapter 10. Product Planning and Marketing. Since technical professionals involved in marketing must have the key role in new product planning and development, this subject is addressed in detail through several analytical questions and answers.

Preparation of sales literature is discussed and examples are shown. Other technical documents include the condensed catalog and engineer-to-engineer applications notes. The role of an annotated price list in product marketing is also defined here.

Chapter 11. Advertising and Sales Promotion. Because advertisements for high-technology products and services must incorporate some new techniques as well as the more traditional methods, this subject is discussed. The typical paid advertisement and the free new-product release are detailed, as well as the preparation required to reap the benefits of such advertising.

The trade show is another excellent way to contact new customers. Techniques and a check list are provided here to help you get the most from each show.

Chapter 12. Your Technical Paper as a Sales Tool. Even though many technical personnel have worthwhile messages for their counterparts at customer sites, they are often reluctant to share these messages through technical papers. This chapter explores the reasons for writing the technical paper, some techniques for writing, and several guides to presentation. All of this has technical impact, but it is also a very important supplement to the marketing program in a high-technology company.

Chapter 13. Sample Proposal. Major excerpts from all sections of a technical proposal are provided, to supplement the case history presented in Chapter 9.

Chapter 14. Sample Proposal Instructions. This is a typical set of proposal instructions from a government procurement document in the $1 million to $5 million range.

To a large degree, each chapter in this book stands alone, providing appropriate information within its subject field. The reader may choose to read through the entire book for familiarization; after that, however, the sections will serve as references when specific needs arise on the job.

WHAT IS A CUSTOMER?

Since the thrust of this book is influencing customers, we should begin with a good understanding of what a customer is, either as a corporate entity or as one or more human beings with personalities much like our own. You must know your customers in order to serve them better.

The question "What is a customer?" seems to have an obvious answer that a customer is the government, a business or an individual who has a problem or need and the money to solve it. But that definition does not go far enough. We need to examine the term "customer" in more detail, to see what characteristics of that corporate or individual buyer may be significant to you in solving the problem or filling the requirement and receiving money in payment for the solution.

The Corporate Customer

Every organization is assumed to have a charter, an assignment, a responsibility. In the private sector, it may be to manufacture automobiles, or to produce and distribute electric power, or to provide data processing services for the federal government. Each specific group within an organization is assumed to have a supporting role in fulfilling the corporate role. (For our examples above—to test automobile alternators before assembly, to monitor the efficiency of a fossil-fueled power generation plant, or to procure and operate a given computer system.) Each individual within each group is assumed to have a specific assignment in support of the group's role—to supervise the group, to purchase equipment and supplies for the group, or to maintain the disk drives.

KNOW YOUR CUSTOMER

- ᔛ *Charter*
- ᔛ *Needs*
- ᔛ *Individual preferences*
- ᔛ *Organization*

In government, functional groups and individual responsibilities relate to overall charters in the same manner. A group's assignment might be to process social security records, or analyze chemicals for a law enforcement agency, or launch a spacecraft. Individual assignments may include data entry into a computer, or maintenance of a mass spectrometer, or analysis of telemetry data.

In order to serve a customer's needs at any given time, one must develop a rapport with that customer to the greatest extent practical. The seller must understand the overall corporate charter, develop a feel for the needs of the groups within the organization, and see how the individuals relate to each other in accomplishing their overall purpose. When offering to sell them a product or service, this is effectively proposing to "become a part of the team." One can't become a team member without some knowledge of what the team is doing.

The prospect of "intelligence-gathering" makes some people uncomfortable, but when they try it they are surprised at how much cooperation they receive from the customer (precluding any security regulations that prohibit sharing such information). People like to talk about what they are doing if they feel that someone else is interested and may be able to help them improve their performance. Corporate policy, in both industry and government, is usually to share information that a prospective vendor should know.

So why can't a visitor get past the lobby? This is simply an institutional way of saying, "We have work to do. We will notify you when we have a requirement." So the seller should use whatever opportunities are available to get to know these potential customers—meet them at trade shows, take them literature, help them to get technical manuals and details on equipment you delivered earlier. The experienced sales engineer can find ways the amateur never dreamed of using.

For any potential customer, the seller should learn, as a minimum, the name of the company or agency, division, charter, location of the company or agency, and the division relationship of this group to other companies, divisions, or agencies, your company's history of dealing with this corporate customer, the organization chart, and the hopes, fears, and biases of the group.

The Individual Customer

Recognize that a customer has the same limitations that beset the seller. Any customer wants the best. Don't we all? However, a budget restriction often prohibits this. Isn't that true for us all? A buyer doesn't want to hear the news that expectations exceed the budget. Does anyone? Quite likely, one finds it difficult to define requirements adequately on paper; so does your customer. For this reason, the "perfect" procurement specification is about as common as the "perfect" technical proposal. The person who is unable to write an airtight specification tends to be skeptical of those who respond, fearing that they have found loopholes and may offer inferior products. The typical customer is looking with suspicion at many potential suppliers, and there is an inherent degree of distrust. Does the listener believe all that is said on a used car lot? No—and the customer doesn't believe all he or she hears in the procurement lobby.

Let's look also at the power of emotional appeal. Engineers profess that they make logical rather than emotional decisions, but emotion plays an inevitable part in the business life of every engineer. The seller should be aware of four strong emotional desires in every human being:

1. Everyone wants to survive. The fear of making the one wrong decision that will cost his or her job, or at least career goals, is very strong in most customers.

2. Many people want a "romance": a new, exciting, pleasant experience. Buying equipment just like other folks bought and using it the same way they did is dull.

3. Everyone wants to get maximum benefit for the money. There is a limited supply, and the strong emotional drive demands a good feeling in return for spending that supply.

4. Everyone wants to be recognized positively. The ultimate pleasure in buying a piece of equipment or a system is to have fellow workers and the boss express praise for the quality of the decisions one makes.

Let's consider the old platitude, "Tell the customers what they want to hear." This makes the sales manager smile, but drives the project engineer up the wall! How can one tell customers the words that warm their hearts while still protecting the company from unfavorable contract wording? There is a way—almost.

Customers want to hear that the proposer can provide a product or system that will "do everything." If they can't afford "everything," say "Yes, we can provide your system, exactly as you have specified it, but in case the price is over your budget, we can recommend some reductions in scope that will save you money, yet not reduce performance significantly." That sounds better than, "I'm sorry, but you can't get what you want for the amount of money you have budgeted." The first is positive and the second negative. Always look for positive wording for the bad news that must occasionally be broken.

In recognizing that customers are driven by emotions as well as by logic, it is wise to use emotional appeal as well as logic in personal contact with each customer and in written submittals. For a particular customer, what can be offered that gives the greatest assurance of survival, romance, value for the money, and recognition?

Each salesperson has a unique way of judging customers. Everyone who relates in any way to product planning, advertising, proposal writing, system engineering, training, documentation, or verbal presentations, benefits from those evaluations. What makes a given customer tick? Know, and grow!

HYPOTHETICAL EXAMPLES

Two hypothetical companies, DACCO and WACKY, are used as examples in this book.

Example 1-1 DACCO. Throughout this book illustrations are used for clarity. Many pertain to Data Acquisition Company, Inc. (DACCO), of Electronics Valley, Colorado, a hypothetical company of 400 employees. It was founded in 1965, and is presently at a level of $35 million annual sales.

The main products at DACCO are high-speed data acquisition assemblies for use with general purpose digital computers. A typical assembly sells for $20,000.

Since DACCO customers are not necessarily adept at system design, the company has established a systems group that buys computers and integrates them into turnkey systems. A typical microcomputer system sells for $200,000; a minicomputer system sells for $2 million. Most of the systems are built around the popular, also hypothetical, micro/minicomputer manufacturer, Clever Computers, Inc. (CCI).

DACCO has two major competitors. Monster Systems, Inc.: 20,000 employees, $1.5 billion annual sales (data acquisition accounts for about 5 percent of their total volume, and sales are based on reputation, not price), and Orientronics, Ltd., Taiwan (200 employees, with low prices but a poor service organization).

Example 1-2 WACKY. The company that gets the blame for mistakes in these examples is Worldwide Avionics Company of Kentucky (WACKY).

FURTHER ACTIVITIES

Each section in this book includes one or more further activities for the reader, who is encouraged to respond to these hypothetical situations to relate the material from the chapter to his or her daily activities. These activities are of great value for groups of two to twenty persons working together in a study of marketing techniques. In that situation, the group should be divided into two or more sections so that opposing roles can be enacted.

Example 1-3 WACKY. The WACKY Corporation has a would-be psychologist on the engineering staff who is unwilling to accept the possibility that a customer (any customer) is a normal human being and who is always looking for hidden meaning in the customer's words and actions. For example:

1. The customer says, "The equipment must operate on commercial-grade electrical power." The psychologist says, "Watch out for that one! He may have the worst power line in the country!"

2. The customer says, "The output must be at least 10 volts." The psychologist says, "Oooh boy! They didn't mention the load resistance, but I'll bet it's one ohm!"

3. The customer says, "The unit must be packaged for commercial air transportation." The psychologist says, "I know what they're thinking! They'll drop the package 10 feet to test it!"

Give the customer the benefit of a doubt. Ask about poorly defined characteristics or make logical assumptions and define them. Remember, the customer really is a normal human being!

2

Bid Decisions and Marketing Strategy

GOOD TECHNICAL

proposals can mean success, and poor proposals usually mean failure for your company.

SALES PROPOSALS AND SUCCESS—OR FAILURE

At some point in the future, a company will hang one of two signs on the front door: *Help Wanted,* or *Out of Business.*

Proper bid decisions and good technical proposals will determine which sign is used. Good bid decisions and proposals mean success; poor bid decisions and proposals mean failure. Is anything more significant to a company?

A proposal reveals the details of an offering to a potential customer. It is usually submitted in competition with the offers from several other companies. It will be judged on its technical content, the capability of the company, and the price. It will be your final and most complete opportunity to present a case. A decision to buy will be made by the customer after all proposals are evaluated. Does anyone dare to exert less than the best possible effort?

Any company can increase its capture ratio on proposals without increasing cost. It can improve the quality of its proposals. The results are up to the workers!

Proposals are written for customers. This may belabor the obvious, but it merits repetition. Too often, proposals are written from the proposer's point of view, with little regard for the customer's desires and needs. Or the proposer may have only a vague picture of these possible customers, who they are, what they do, how they operate, and what they really need. The first principle is, therefore, *get to know the customer and know what is wanted before the procurement comes out.*

The customer defines his or her needs in a package of documents called a "bid request." Items to which the engineer must respond are in the "statement of work." The ideal bid request is a clear statement of the customer's requirements that can be answered by a clear proposal. Such a request sets forth performance specifications, materials, workmanship standards, reporting requirements, schedules, and other needs. Each of these factors becomes a topic in the proposal and is answered to the extent necessary to satisfy the customer. The ideal bid request is, however, no more common than the ideal proposal. If a customer has not stated the requirements clearly, the proposal manager must determine what the customer needs and define those needs for the proposal team.

The more the seller knows about the customers, the better chance the proposals have.

THE BID DECISION

If you aren't prepared and you aren't willing to go all out, don't bid the job!

The first step in creating a successful proposal is a proper choice of bid opportunities. The capture ratio will be much higher under some conditions than under others. A calm and rational analysis of the conditions is necessary to select the right bid sets for which to prepare proposals.

A bid meeting should be held as soon as is practical after the customer's specifications have been received. If sufficient advance information is available, an earlier meeting may be practical. It should be chaired by the general manager or marketing manager and be attended by a high level representative from each department that will be contributing to a logical analysis of the bid set.

In preparing for the bid meeting, the technical professional who is assigned should collect all the available facts and prepare a handout for each attendee. The handout should include a simplified functional sketch or diagram of the desired equipment or a brief statement of the desired services, a rough estimate of the cost of the procurement, a rough breakdown of these costs (standard products, materials, engineering labor, software, assembly labor, and so forth), the due dates of the proposal and the equipment, and an estimate of the proposal effort in person-weeks.

The responsible technical professional should present the significant facts:

1. Who is the customer?
2. Does the customer have money?
3. What is our history of dealings with this customer?
4. What is our background on this particular procurement activity?
5. Who are our competitors? (What are their strong points? Their weak points?)
6. What must we do to have the best chance for success?
7. If we do these things, what are our chances?

If the chairman decides to pursue the bid, the company management should appoint a *proposal manager* immediately and establish a date and time for the kickoff meeting. From this point forward, the proposal manager must have the full and open support of the general manager and all department managers. Anyone in the company who learns of significant new facts that might make it advisable to reverse the bid decision should present them immediately to the proposal manager for reconsideration and possible immediate reconvening of the bid meeting. The bid decision should never be reversed by anyone except the person who made the decision, nor should any phase of proposal activity be suspended or even delayed in anticipation of a reversal of the bid decision.

Do you want to win? Then go all out!

- ✍ *Put a good person in charge*
- ✍ *Give that person a good team*
- ✍ *Prepare, prepare, prepare*
- ✍ *Work with the customer*
- ✍ *Gather and use information*
- ✍ *Establish a strategy*
- ✍ *Review, review, review*
- ✍ *Squeeze the cost estimates*

As The Gambler says, "You've got to know when to hold and when to fold!" More profit can be lost on one bad bid decision than on a dozen bad decisions not to bid.

WACKY Was Not Prepared

Example 2-1. Our friends at WACKY saw a notice in a week-old U.S. government *Commerce Business Daily* regarding procurement of a system similar to those built at WACKY. They requested a copy of the bid set and received it in a few days. The due date was only 15 days away when they made the bid decision.

Old timers say, "Don't bid a job if your first knowledge is in the *Commerce Business Daily*." Nevertheless, WACKY dropped everything else to staff a proposal team. They did not know the customer agency, were late in starting, and were unable to offer anything unique for the bid. They nevertheless charged into the proposal effort, jeopardizing a few other opportunities along the way and, of course, they lost.

Instead of the WACKY way, a company should spend its proposal effort on opportunities it has been tracking. It is not impossible to win from a standing start, but it is extremely unlikely.

STRATEGY FOR A WIN

Early in dealings with a potential customer, long before any of the specification has been written, the seller should begin to develop a "win strategy." Many decisions which are postponed until the critical bid package has arrived should have been considered months earlier. What *must* we do to win? What *are* we doing to win? Does it involve development of a new piece of equipment? Should we write some new software? Would the customer like to see one of our systems at an operating site? Can we team up with another company to strengthen our position? Do we need to hire a consultant to fill in a weak area in hardware or software? *Push the customer's button! Solve his "worry" problem!*

YOUR PROPOSAL BUDGET

Good business practices require that management establish a budget for the preparation of each proposal. This will usually be related to the expected size of the procurement. The company or division should evaluate the allocations frequently, revising them to reflect the realities of the competition. For each proposal, a budget should be established to reflect the unique characteristics of the opportunity it presents. Extra money might be worth allocating to a proposal effort that represents attractive new market opportunities and less money to a routine proposal that represents no new opportunity.

The budget, however, should never become a noose around the necks of the proposal team members. The proposal manager should use it to establish the team organization and define the days of participation by each member, but no one should use it as a club over another's head in the heat of the proposal activity. A few trimmings here and there may mean the difference between success and failure.

SMALL, MEDIUM, AND LARGE PROPOSALS

A company should be capable of handling both large and small proposals within the same basic framework. A small proposal may involve just the proposal manager for only a few days. The largest proposal may involve more than a hundred people for 3 months or longer. The same basic principles apply to small, medium, and large proposals.

PROPOSAL PLAN CHECKLIST

☐ Did work on this proposal begin long before we received the request for proposal (RFP)?

☐ Is the proposal a sales tool—the end, not the start, of our marketing effort?

☐ Do we understand the customer's problem and will we clearly state our understanding in the proposal?

☐ Is the proposal, like any business endeavor, planned and organized in advance?

☐ Will we make an objective review of our proposal before it is submitted?

☐ Will the proposal present a persuasive, often repeated, theme throughout its text?

☐ Does the proposal involve more than just a good technical approach?

☐ Does the executive summary, on its own, tell our story?

☐ Will the technical approach discuss requirements, alternatives, and rationales for proposed solution?

☐ Are the program and management plans we will use feasible, specific, and comprehensive?

☐ Are our company's qualifications, and the resumes which we offer, germane to the proposed effort?

☐ Does the proposal outline address every requirement in the RFP?

☐ Can the customer/evaluator read, understand, and believe our proposal without confusion?

☐ Is each volume and major section of our proposal to be summarized at the start?

☐ Did we take time and care in preparing the statement of work and work breakdown structure?

☐ Will we make maximum use of illustrations, tables, and charts to convey information?

DEVELOPING THE SUBJECT—SOME EXERCISES

1. You are an employee of DACCO. Your company has been asked to bid on a data acquisition system to be purchased by a prime contractor to the U.S. government. It will be used to collect data at a wind tunnel test facility. Your known competitors are Monster Systems and Orientronics. You have never bid on a system this large. You feel that your price could be lower than Monster Systems but higher than Orientronics. What techniques can you use to develop a win strategy?

If you are in a class, assign one third of the class to DACCO, one third to Monster Systems, and one third to Orientronics. Develop your group strategies separately and then critique those of the other groups.

2. You are the manager of a hardware project team at DACCO. Your team purchases Winchester disks and streamer tapes for integration into systems—in other words, you are someone's potential customer. Create a profile of yourself in this role suitable for use by another company in understanding how to deal with you.

3. You are preparing a DACCO proposal for submission to Octane Oil Company. They will use your system or that of a competitor to measure the heating effect of petroleum products. Create a profile of the decision-making group at Octane Oil, a detailed profile your sales engineer in Texas would make as you start your proposal activity. If you are in a group, share the results of these exercises with others. What good information did you include? What did you omit?

3

Getting Started

PROPOSAL MANAGER URING THE PREPARATION of a proposal, one of the most important persons in the company is the *proposal manager*. Proposal management demands the very best of management capability, that is, it does not allow one the luxury of a postponed decision, a slipped delivery, a Saturday on the golf course, or a closed office door. While the proposal is being prepared, the proposal manager must have the latitude and authority approaching that of the general manager. He or she must be able to cross departmental lines, authorize expenditures (within the budget), and overrule anyone in the company except the general manager.

The proposal manager must be chosen for ability and expertise in management, diplomatic marketing strategy, overall understanding of the company capabilities, and positive attitude. This person may be an engineer, a marketing professional, or an administrator. He or she need not understand the details of the proposed equipment, as long as he or she understands the generalities of the equipment and its application.

Finally, and most important, the proposal manager must have contagious optimism. Here, as almost nowhere else, optimism shows in the final product. Seldom does a pessimist exceed his expectations; the optimist often meets his highest hopes.

A Sad Story at WACKY

Example 3-1. WACKY was preparing a major proposal, its largest ever, for a very important program. The proposal was so important that the division manager served as the proposal manager (his first such assignment). He managed the proposal as a drill sergeant manages a group of new recruits. His normal duties suffered as a result of his involvement on the proposal and the proposal suffered as a result of his management. WACKY lost the job.

Division managers should manage divisions, and proposal managers should have their complete, obvious, continual support. This is the way to manage divisions effectively and win contracts.

THE PROPOSAL TEAM

Set up an effective proposal procedure, organize it, schedule it, and give it top priority. It's a fact of life in high-tech industries that proposals and pressure are synonymous, because: a) Proposal making is a complex effort of coordinating many skills such as engineering, legal, accounting, purchasing, technical writing, management, and printing; b) It's a task added on top of the normal workload; c) And then there's that deadline!

There will always be pressure on the proposal team. In moderation, pressure is a fine stimulant to getting the job done (and a fine brain stimulator, too). But runaway pressure can cause oversights, inconsistencies, and even gross errors. Worse, it can dishearten the team so that ensuing proposals become progressively less effective.

Pressure relief requires positive thinking team members and good management of the proposal effort from start to finish.

Set Up an Effective Procedure

Preparation of any proposal follows a preordained sequence. The manager must set up a standard operating procedure based on this chain of events. Then all participants know what must be done, how it will be done, who will do it, and when it must be completed.

Organize It

Use the basic tool—the outline. Work assignments can be made from it and all participants are provided with a map to guide them over the route.

Schedule It

Set up a realistic schedule for the proposal and adhere to it. It's human nature that some people fall under the whip of the deadline before they begin to produce. If they only contend with the final deadline, their single tardy burst of energy is too late and chaos results. Management's role is to set up a series of subdeadlines and enforce adherence in order to pace the effort. Overtime is then predicted and distributed over the entire period rather than round-the-clock just before the deadline.

The schedule must allow time for a review of the final manuscript. This crucial step prevents any late-sprouting errors from making it through to the customer. This time must be used also to make the proposal read as though it were written by a single author. It must "flow." A proposal that reads as if it were written by committee members who never met each other is not a winner.

Cursed as the deadline is, it's the proposal manager's best ally, as it forces people to make decisions that otherwise are put off interminably. Is anyone ever "ready for the holidays"? Of course not, but we know it will never be postponed, so we deal with the deadline.

Give It Top Priority

No obstacle must interrupt the chain of events that constitute proposal writing, from inception to printing and binding. Roadblocks often crop up in the strangest places. For that reason, top management must establish a blanket priority.

It is a fact that good management and effective procedures will improve the company's proposals and enable it to produce more for the same dollar expenditure, resulting in more business per dollar spent on the bid and proposal. Most important, it reduces the strain on the valuable proposal team, for which they will be grateful.

We must recognize that some bids are not large enough to require a large team and, in fact, the smaller proposals may be written by just one or two qualified persons. Whatever the size of the proposal or the team, the same general principles apply.

WORK BREAKDOWN STRUCTURE

An excellent document for managing and organizing the bid effort is the work breakdown structure (WBS), which helps to define the elements of a technical response. It is a necessity in determining costs of a job and is required by many agencies as a part of the bid submitted.

The WBS is a top-down organization of the proposed effort, showing in separate branches the major elements. The proposal manager should generate this work breakdown structure to remind each estimator of the tasks to be estimated; it will serve as a central collection point for the accountant who collects and assembles the estimates; it will allow the proposal manager, in the last-minute panic of the pricing decision, to see exactly where the costs are located; it will enable negotiators to deal with the customer's audit team if there is such an activity; and it will serve as the foundation for a program budget when the contract is received.

A WBS may be very simple and brief, or may consist of more than a thousand items, depending on the needs of its users. The only criteria on the depth of detail is that it be suitable for these needs and that the delineation between recurring and nonrecurring tasks be clear. One simple WBS for a Data Analysis System might look like this:

1.0	Data Analysis System
1.1	Box A
1.1.1	Design, nonrecurring
1.1.2	Construction, recurring

1.1.3   Test, recurring
1.2   Box B
1.2.1   Design, nonrecurring
1.2.2   Construction, recurring
1.2.3   Test, recurring
1.3   System computer
1.3.1   Selection, nonrecurring
1.3.2   Purchase, recurring
1.4   System design, assembly, and test
1.4.1   Hardware design
1.4.2   Software design
1.4.3   Assembly, recurring
1.4.4   Preparation of test procedure, nonrecurring
1.4.5   Test, recurring
1.4.6   Preparation for shipment, recurring
1.4.7   Shipment, recurring
1.5   Technical documentation and training
1.5.1   Preparation of technical manuals
1.5.2   Preparation of training course
1.5.3   Publication of one set of technical manuals
1.5.4   Training, one week

The WBS outline should be generated early in the proposal process and should be sufficiently flexible to accommodate things which are learned during the proposal sequence. No one except the proposal manager should be allowed to change the WBS. This is not to imply that changes are to be avoided, but rather to emphasize that there must be orderly control of the document.

YOU VERSUS YOUR COMPETITORS—
THOSE SIGNIFICANT DIFFERENCES

Every competitive procurement involves one or more significant differences between a bidder and their competitor(s). If it doesn't, it's up to the bidder to create them. If they can't, why did they decide to bid? The proposal manager must recognize these differences ("discriminators"), and take action to capitalize on the advantages over the competition while neutralizing the disadvantages.

Discriminators come in three categories:

Aha!: You are significantly better than the competition on this item.

Oh No!: The competition is significantly better than you are on this item.

Ho Hum: This item difference doesn't really matter to the customer.

Don't ignore or downplay the reality of the situation. Improve the company's position in the eyes of the customer by having the proposal team:

☑ Brainstorm their way through an enumeration of the discriminators on the procurement,

☑ Carefully plan an approach to capitalize on the *Aha's* and neutralize the *Oh No's* (by blowing your horn, and not running the competitor down),

☑ Prepare an executive summary that builds up the case and then carries the idea through to each section of the proposal.

Some of the factors that can be classified into the *Aha* or *Oh No* categories include: 1. Availability of standard hardware and/or software to do the job. 2. Experience with this particular type equipment or system. 3. Existing equipment at the customer location where this equipment or system will be used. 4. Past experience with the same customer. 5. Availability of a critical piece of equipment which competitors can't match. 6. Lower cost. 7. Better field service.

This listing of discriminators and the strategy to handle them is essential if you are to make best use of your resources in writing the proposal.

KICKOFF MEETING

Within one or two days after the bid set arrives, the proposal manager must hold a *Kickoff Meeting*, with all the proposal team present. The kickoff meeting sets the tone of your proposal. At this meeting, the general ground rules of proposal preparation are set forth and everyone is assigned specific tasks.

Determining the Plan of Attack

Preparation for the kickoff meeting is likely to be the most critical period in the proposal effort. The proposal manager must decide the basic plan of attack: Propose in exact conformance with the requirements? Propose a more exotic approach? Take exceptions to reduce the cost?

Be terribly careful about those exceptions; on many government procurements, taking exceptions when one of the other bidders is fully compliant will lose the job.

What is the proposal outline? How many illustrations will be used and what will they show? The sales engineer will be able to supply some of the recommendations, but the proposal manager must make the final decision on each matter.

Defining the Customer

At the kickoff meeting, the proposal manager must describe the customer's requirements and the company's plan of attack. This person (or the sales engineer) will describe the customer—application, strengths and weaknesses, and personality. In the kickoff meeting, you should determine the ways in which you can appeal to the basic emotions of the customer, that is, the fear of making a mistake, or the desire to be a hero, or whatever else motivates or frightens him or her. The competition should be analyzed as it was in the bid meeting. Copies of the simplified block diagram of the proposed bid and other descriptive sketches should be distributed. Each team member should get a copy of the portion of the bid set applicable to his or her task. On larger proposals, a special work area ("war room") should be set up where any team member can examine any document from the complete bid set at any time.

Use Consistent Terminology

The proposal manager should determine the glossary of terms to be used in the proposal. Nothing detracts from a proposal more than the use of two or three terms by different contributors to describe the same portion of the equipment. If the customer uses certain terminology, the responder must use the same terminology. Even the name of the company should be standardized.

Proposal Outline

The proposal outline should show the approximate number of pages in each section and the person responsible for each section. The proposal will lose much of its impact if the writer spends too much time on one section and short-changes another section. In some proposals an ordinary operation is explained in boring detail but only a half page is allocated to the manufacture of 100 sets of equipment. What do you suppose the manufacturing manager at the customer's plant thinks of that?

Scheduling Time and Apportioning Budget

A detailed working schedule should be derived at the kickoff meeting. Know the customer's due date. Work back from that to determine the mailing date, the date for start of printing and assembly, the date for giving longhand written versions to the word processing personnel and sketches to the graphics department, the date for submission of typed draft to the proposal manager for review, and so on through all the subtasks which are involved.

Work backward to devise a schedule for pricing activity and don't forget the interdepartmental due dates, such as, when will manufacturing get sufficient information for their pricing effort? When will the reliability engineer get a parts list for reliability calculations? When will the technical illustrators get sketches from which to prepare the artwork required for the proposal?

The newspaper was on the front steps of subscribers this morning because its various departments met their staggered deadlines. Some of them supplied modifications after their initial submittals, but a modification is simpler to handle than a late submittal. Not a single reporter was fully satisfied with his or her input and every article would have been better if subscribers had been willing to wait until tomorrow to read today's news! No one on a proposal team will be satisfied with his or her input either, and every section would be better if there were "just a little more time." The adults will meet their deadlines though, and the children will whine.

Successful completion of a last-minute fire drill does not reflect favorably on management capability. Good management avoids a drill in the first place.

STORYBOARDING

Storyboarding can help to produce a better proposal in a shorter time with less headaches.

How It Works

An interesting way of planning the proposal is to imagine that one has already built the equipment the customer needs, and invited the customer to the plant to see his equipment in operation. First, the offerer calls to explain briefly the four or five main reasons why he should come to see the equipment (the proposal synopsis). The offerer sends him the planned agenda for his visit (table of contents). On the short ride in from the airport, he is told of the suitability of the equipment to his needs and the reason for the enthusiasm about the equipment (introduction).

When they walk into the room housing the equipment, they stand several feet away from the device and look at its general features: its size, how it interfaces with other equipment, its simple operation, ease of maintenance, and the main advantages of this approach over those which were rejected (general description).

The host shows him a simplified block diagram, and explains the functions of the equipment, looking at some of the more detailed diagrams to illustrate the unique features of this particular design (technical description). He is told why a specific type of component, or packaging, or power supply was chosen. They spend a few minutes discussing the operating manual and the maintenance manual, the spare parts list, and other documentation, and how the company installs and maintains the equipment (documentation and services).

Obviously the offerer wants the customer to meet the key engineering, manufacturing, and support people on the job, to see how they worked together and to understand how top management monitored the job and supplied direction and support.

They want him to meet any outside consultants, major vendors, and anyone else whose contribution to this new purchase was significant (project organization). They are proud of the orderly manner in which the project was designed and constructed (schedule and work statement).

The host goes on to show the customer the tabulations of the specifications for the equipment and how these match his requirements. In cases where they don't, an explanation is made of how this approach is actually better for him because of price, performance, reliability, or some other reason (specification analysis).

Finally, the customer is given a package of material to read on his plane trip home. This will include any applicable technical reports, detailed reliability analyses, a history of the company and some details of its financial capability, a list of related jobs the company has completed, and other backup information deemed necessary to satisfying his needs (appendix).

This overview can serve as a springboard for "storyboarding" on the large proposal, a technique that has been proven to:

- Elicit helpful comments from team members
- Highlight your important selling points
- Bring consistency in your theme statements
- Eliminate major rewriting
- Support and organize your graphics
- Make your proposal easier to evaluate
- Identify optional additions
- Save time and money

How You Do It

At the beginning of the proposal schedule a one-page planning document or "storyboard" is created for each numbered paragraph in the proposal. Each storyboard is reviewed and revised by the proposal manager as necessary and then becomes the basis for preparation of each paragraph. Figure 3.1 on the next page shows the layout of a storyboard page.

When this approach is used, it is important that the planning be done early, the storyboards be filled out early, the re-

STORYBOARD # _____

Proposal Paragraph: _____ Subject: _____

RFP Reference: _____ Subject: _____

Requirement: _____

_____

_____

Our Bid Strategy: _____

_____

_____

Theme Sentence for this Paragraph: _____

_____

_____

Sub-heading and Emphasis	Illustrations

Page Budget: __ (double-spaced). Figures: ____ Tables: _____

Comments: _____

_____

_____

Preparer: _____ Phone: _____

Figure 3.1   Layout of a storyboard page

views of all storyboards be conducted with top priority by the proposal manager and his or her designees, and that all writers conform to their storyboard plans while writing the relevant proposal paragraphs.

Example 3-2. WACKY's proposal team was working on a bid set from the U.S. government that included references to some relatively minor military specifications. The proposal manager had made the decision to take exception to these specifications, being sure that the proposal would not be declared nonresponsive because of these exceptions.

A department manager was told of this decision by a member of the proposal team. Upon reading the procurement specification, the department manager judged that, despite the proposal manager's instructions to the contrary, these specifications were to be taken literally. The department manager decided that the company should not bid because of these specifications. But instead of telling the proposal manager, the department manager told his staff to stop working on the proposal. This created no end of havoc before word got back to the proposal manager and general manager.

The general manager restored some order to the proposal team, but the near disaster situation occurred because one manager did not follow the most elementary rule in proposals: "Only the person who made the bid decision can reverse it."

Take all doubts to the proposal manager. Ask for a review of the bid decision if you wish. But keep on working by the proposal manager's rules until or unless the bid decision is reversed.

OUTLINE

The outline forms a framework on which all team members can build. Make it serve a good purpose. A proposal is organized by its outline. A good outline from the proposal manager is vital to its effectiveness. The purpose of an outline is to:

☑ "Build in" responsiveness by ensuring that all customer requirements are answered.

☑ Guide the team in preparing the proposal.

☑ Facilitate the process of writing the proposal.

Sources

Topics in the outline are drawn from two sources. The primary source is, of course, the customer's statement of work. Requirements are slotted into the outline first, to ensure that key points will not be overlooked.

The second source is the writer, who originates topics that supplement and expand upon those supplied by the customer. Thus begins the creative process that will culminate in a proposal.

How to Set Up an Outline

The proposal manager should build the outline by following these steps:

1. Read the statement of work.

2. Set up major sections of the outline.

3. Study the statement of work and make note of all topics to which the proposal must respond.

4. Supplement customer topics with those of one's own choosing, selecting those that explain and justify the approach and clearly indicate the program.

5. Determine the weight of each topic, whether of equal weight or subordinate, and then decide on their order of presentation.

6. Assign topic numbers, using subsections within each topic: Section 1, and its subparts 1.1, 1.1.2, etc.

What to Consider in Setting Up an Outline

Some of the major considerations to be worked into the outline of the technical proposal are presented below:

1. A description of novel ideas or technical approaches developed in the analysis of the problem.

2. A statement of the major technical problems that must be solved, with an indication of the amount of effort budgeted to

each. This is a major check point for the customer in both technical and cost areas, since it shows him whether or not the writers truly understand the problems inherent in the procurement.

3. A discussion of the technical approaches that have been explored or will be explored, and why this specific approach can be expected to yield the desired results.

4. A brief discussion of alternative routine or speculative solutions that have been explored and rejected and the reasons they were rejected. This assures the customer that the engineers have not come to a hasty resolution of this problem and highlights the extent of the research undertaken to arrive at this solution. It also subtly plays down approaches that may be used by your competitors in the other proposals your customer will receive.

5. Identify unrealistic and unreasonable performance requirements and associated costs. The customer may not always realize the effect of some of the performance requirements imposed by the request for proposal, or he may not be aware of the delay and cost associated with accomplishing them. These areas should be pointed out in the proposal and possible alternative solutions presented that involve less time or lower cost. Identify the more difficult areas or problems to be solved and detail how a breakthrough in the state of the art will be achieved to meet the performance requirements.

6. State where you intend to deviate from the specifications. Be careful here because the customer may resent any deviations and interpret them as a reflection on his understanding of the problem. Deviations should be kept to a minimum and adequate justification provided. Don't forget: this customer hasn't bought a system or written a system spec in 5 years, or maybe 10.

7. Show that the offerers did their utmost to use existing items or components. If new components must be developed, explain why existing ones cannot be used.

8. Show the relationship of the proposed contract effort to any existing or previous contracts you have fulfilled for this or other customers. Indicate the customer, the type of project, the

funds available or already spent, and the results achieved to date. This is an especially important point because if the company is already engaged in effort required by the contract, the customer will receive the benefits of the work they have done previously.

9. Describe the technical services to be provided, including site operation and maintenance, field support, provision of spare parts, systems analysis, and off-site operations.

10. Provide resumes of the technical personnel who will satisfy the requirements of the request.

Tailor the Proposal to the Customer

Here are some ways to make it easy for a customer to evaluate a proposal and to do business with the offerer:

1. Propose what the customer has asked for, not what the proposal submitter thinks is really needed.

2. Assign "customer-compatible people" to the proposal. Visualize the proposal writers engaged in a lively discussion with the customer's counterparts; chances are they'll make a good match.

3. Propose the minimum solution that will satisfy the customer's requirement. People can't buy Ferraris on Volkswagen budgets.

4. Make the proposal clear and logical. Write it well. The customer will respond with his or her business.

5. Respect the customer. This is an element often overlooked, and if you don't, the writing will give you away.

6. It's a hard job writing a winning proposal, but don't be grim about it. Enthusiasm on your part can awaken a like enthusiasm for your product in the customer.

7. Be specific, not redundant. Don't wash the customer away in a sea of words.

8. Remember when writing for technical people not to take the advertising man's approach. Let hard facts do the hard selling.

One of the customer's chief concerns is that the bidder understand the requirement. Again and again, proposals fail to demonstrate that understanding. While a writer may truly understand the customer's problems or needs, often that understanding is not demonstrated clearly in the proposal. One extremely common mistake (in spite of customer admonitions to avoid it) is using the customer's own words to describe the problem or requirement. Obviously, repeating the customer's own words is not proof of understanding.

At this point, we recognize the fact that smaller proposals may be written by just one or two qualified persons. But, whatever the size, the same general principles apply.

To Win:

☑ *Propose what the customer has asked for.*
☑ *Propose the minimum solution that will satisfy the customer's requirement.*
☑ *Assign customer-compatible people.*
☑ *Make the proposal clear and logical.*
☑ *Respect the customer.*
☑ *Maintain enthusiasm.*

SUGGESTIONS FOR REVIEW

1. Look back at the exercise involving DACCO, Monster Electronics, and Orientronics, and the proposal for the wind tunnel test facility in Chapter 2. From what you are told and what you can assume about the situation, define each *Aha!*, *Oh No!*, and *Ho Hum* for DACCO.

Within a group, assign one third of the members to DACCO, one third to Monster Electronics, and one third to Orientronics. Compare and critique each group's lists.

2. You are the proposal manager for DACCO. Build a proposal outline for the wind tunnel test facility system. Use your imagination. Read between the lines and speculate on the best way to present your company's capabilities.

In a group, assign two or three teams to devise a proposal outline, and then compare and critique each other's outlines.

4

Your Proposal—
The Technical Parts

Establish and maintain
*a proposal resource file. It makes life much easier for the proposal
writers.*

PROPOSAL RESOURCE FILE

One of the most frightening sights to some engineers or programmers is a blank piece of paper! "What do I write?" "How should I start?"

To ease the burden of facing a blank page, writers establish a proposal resource file in each group that contributes to the proposal effort. This file must be indexed logically and kept in a loose-leaf notebook to invite the addition (*but not the removal!*) of information. Since the word processing group will prepare most of the information from which the file is derived, be sure to write the original document number on the first page of each entry in the file and save the page numbers from those original documents. Don't overlook the value of sketches from earlier proposals as well. Again, save the identification number from which the original of each sketch can be retrieved. Many of the sketches will be reusable as is, and even if a change must made, it is easier than starting over again.

The front page of a resource file should list every proposal which has been searched for relevant excerpts. This ensures that repetitive searches of the same proposal will not be made.

Be careful when "recycling" material from previous proposals. Although it is a time-saver, avoid the ever-present risk that the original proposal material referenced an earlier customer by name or by unique application. Mark those specific references on the resource material when it is filed, in order to safeguard against leaving the "U.S. Army" name in a new proposal for the "U.S. Air Force."

WRITING THE ABSTRACT

Look at your win strategy before writing the abstract. The first page should be an abstract or "executive summary" of the proposal. Identify the four or five strong points in the offering and bring them out forcefully. Be specific. Instead of saying, "The system has high reliability," say, "The mean time between failures for the system is 2370 hours."

What does the customer want to hear? If the writer can say it truthfully, it should be in the abstract. If not, save the bad news for later.

Look at the win strategy and those significant differences before writing the abstract. Then write a *draft* copy and distribute it to all team members at the kickoff meeting and to the marketing manager, product line manager, and general manager for detailed comments. The first draft will not be suitable as a final version so prepare it early enough to stimulate comments from all who are familiar with the bid set and/or customer. Keep modifying the abstract and keep accepting comments. The abstract should be started early in the proposal effort but should not be finalized until the last piece of the proposal is typed and assembled.

WRITING THE INTRODUCTION

Start the proposal with some well-chosen words. Here is the writer's chance to sell the equipment—or to bore the customer. A well-worded introduction will show the customer, in a brief statement, the essence of the bid. Show, in broad terms, how the company has solved similar problems for other customers. Show the general approach to the solution. Remember those "significant differences?" Be sure to accentuate the positive points here.

If the section leaves out any important advantage of the offering, the introduction is inadequate. The engineer who wrote the specification will read the entire proposal, but his or her boss will read only the interesting parts. *Be sure that you make the sale in the abstract or executive summary and the introduction!*

The following section highlights some introductory topics that can be considered typical. Although many topics are presented, only certain ones are applicable to any one proposal. Also, certain topics seem redundant. All possible topics are included, however, because the value of the list lies in the wide choice available. The list is essentially a memory jogger designed to ensure that vital topics are not overlooked. Depending upon its weight, a topic may be covered in a single sentence, one or more paragraphs, or an entire section.

The introduction should not exceed five pages. Two or three are even better. Although brevity is the keynote, the introduction must not short-change essential points. Although generalities are needed here, these must be informative and point the way to what follows.

When writing introductions, avoid the danger of paraphrasing, or even quoting at length, from the customer's statement of work. The result is a watered-down proposal. Words are precious in selling ideas: don't squander them on nonessentials.

The introduction should highlight (in short, snappy statements) five or six distinct advantages the company offers. These should be emphasized throughout the proposal. The fact that the offerer used the same features in the abstract, and will use them throughout the proposal, is not an issue. Make sure they come through to every reader!

Points to Remember When Writing an Introduction

Show the essence of your proposal; show that you understand the problem; bring out those strong points; be brief (not more than five pages).

Anything that is special about the proposal should be introduced here. If there is some unusual and relevant resource (such as an electronic microscope or remarkable proprietary process) use it to command attention here, at the beginning of your proposal. Make your customer read more eagerly and turn to the second page with anticipation. Such items might include:

1. Any unusual and impressive resource such as a prominent technical or professional authority, an unusual research library, proprietary processes, equipment.

2. The staff's unusual qualifying experience or accomplishments that have direct relevance to the product or service you are selling.

3. A subcontractor who is especially well qualified or can offer any other special resource.

4. A special insight into the requirement or problem that merits discussion here rather than in a section on the requirement itself.

5. A special promise or pledge regarding results or effort, such as an absolute guarantee that the company can carry out the entire project in significantly less time than projected in the work statement or with substantially greater results. This may represent an attention-getting lead on the main strategy if it represents a dramatic offer.

Some typical topics to be addressed in the introduction are:

1. Basis for proposal submittal (response to formal *Request for Proposal*, letter, purchase request, or unsolicited proposal).

2. Customer document number, date, and amendment numbers (if any).

3. Additional information (bidder's conference, date, and location), and how it was obtained (letter, telephone, and so on).

4. The company's identification, and the reasons why it should receive the contract. Include a review of the principal competitive advantages.

5. Objective, scope and duration of the program.

6. Statement of the problem.

7. Alternative solutions. Relationship of proposed program to other in house programs, company-sponsored independent technical effort, and company long-range business objectives.

8. Description of end product.

9. A clear, concise statement of the technical requirements that the proposal fulfills.

10. Description of the expected end result of the program.

11. Relationship of proposed work to the state of the art, including the presently available components, equipment, techniques, or systems.

12. The value of the program for the *immediate* application and a prediction of performance in relation to *future* requirements.

13. Expression of interest in conducting the work.

14. The company's qualifications, including specialized facilities and related management and technical experience.

15. Relationship of proposed program to previous successful programs.

16. Factors that will ensure the operational effectiveness of the end item.

17. The unique program for long-range servicing of the product.

18. A *brief* summary of the information set forth in the preceding items.

19. A *brief* summary of the contents of the remainder of the proposal.

Example 4-1. The proposal manager at WACKY asked one of the team members to write the all important introduction to a proposal. Theorizing that the first day would give him more free time than any other day, this person sat down with a blank sheet of paper to start the job. Unfortunately, at the be-

41

ginning of the proposal effort, the writer knew virtually nothing about the features the team would offer and very little about the ideas that would be generated within the next few weeks. Yet the draftee bravely started—and finished—a very weak introductory section.

Four weeks later, when the proposal's components were being thrown together hurriedly, this introduction became set in concrete, not because anyone liked it but because there was no time for a rewrite.

Remember: Do a *rough draft* introduction first, and hand it out at the kickoff meeting to introduce the proposal's themes. Rewrite it *after* everything else in the proposal is written and critiqued. Save those necessary 2 hours to do it right.

WRITING THE SYSTEM DESCRIPTION

The second section of a proposal defines the system, so that the reader will know that what he or she needs is understood. Don't assume that because someone at the customer agency wrote a system specification everyone there knows exactly what is to be supplied and that an overall description is not needed. (That is an invitation to disaster.)

In these six to ten pages the writer has the opportunity to define the overall offering and sneak in a few selling points. There are several reasons for defining a system up front for the customer:

1. The person or persons who wrote the specification want to be assured that they are understood.

2. The top level managers at the customer agency may not read any further, so this may be the last opportunity to sell them.

3. If you are offering a few features that the customer didn't ask for, and didn't expect for extra cost, make sure the customer learns about them here.

4. Even the people in the company who are not deeply involved in putting the proposal together will benefit from an overview of the system.

A draft of the system description should be given to all proposal team members early in the game, ideally at the kickoff meeting. The system description should be limited to 6 to 10 pages. Within that space, however, the writers have the opportunity to define the overall offering and sneak in a few selling points. Don't miss the opportunity!

A simplified block diagram should be presented with the system description. Emphasize the word "SIMPLIFIED." Keep in mind that the engineer's boss and management people from other areas are not likely to read beyond this point in the proposal. They should be able to understand, through a few simplified illustrations, the approach the company is taking.

This is also the place for the proposed system layout, physical structure, and function.

The "significant differences" must be brought into the system description with the positive differences built up as strong features and the negative differences defended. The storyboard outline must be used as the basis for this section.

Another factor to be considered in the system description is the ease of operation and maintenance. Proposal writers often erroneously emphasize the most complex portions of the black box or system, and most customers cannot understand this emphasis. They do not doubt that you have the ability to build the guts of a black box and they are usually more concerned with the simplicity of operation and capability for maintaining the system. Give ample consideration to these capabilities.

It is surprising how many proposals are submitted with no clear statement of how many pieces of equipment are to be delivered! While this is one basic omission, some more subtle ones relate to the *quantity* of each type of plug-in module, or the amount of *documentation*, or the amount and type of *support services*. Make sure all of these are well defined in this section as they also help to provide a clear definition for the cost estimates.

WRITING THE TECHNICAL DESCRIPTION

The technical description is the engineer-to-engineer communication channel in the proposal. The proposal must show respect for the customer engineers, not insult their intelligence by telling them the obvious. Don't start in the middle of the problem, or they will be lost at the outset. Stand back, take an objective look at the offering, and then proceed with an orderly description of the equipment.

1. How does it operate?
2. Why was this approach chosen?
3. Is it easy to maintain? Why?
4. Is the offerer pushing the state of the art in any areas?
5. What is the probability of success? (If the team doesn't succeed, will it be a minor deviation or a catastrophe?)
6. Do any of the customer's requirements have a particular impact on the price, and if so should an alternative approach be considered to reduce the price?
7. Should other considerations be presented to the customer to help in their analysis?

The storyboard should form the basis for each part of the technical description. Use sketches and charts where they are relevant to simplify and clarify points. Make sure that the technical section is not all electronic or all mechanical. Don't forget to describe the packaging of the equipment. Why was this method chosen? What does it mean to the customer? Lower cost? Or easier maintenance? Or lighter weight? Tell them. Show sketches of the unique features, if any. Let them see how important this phase of the proposal is to them.

Consider what is the outside world's interface to the proposed equipment. If it is an operator, show the control panel and how the operator has adequate (but not excessive) control of the operation. If it is a mechanical device, sketch the interface to assure the customer that the problem is understood. If it is to interface electrically with another device or devices, describe the connectors, voltage levels, and other pertinent details. A few well chosen, simple illustrations make the material easier to understand. Don't forget to use illustrations for the mechanical details, such as module mounting, front panel layout, and cooling.

Crises in proposal preparation inevitably arise from the discovery that the customer's specification is confusing in some areas, but they need not cause panic. The proposal manager may be able to deduce the answers, or he or she may present the questions to the customer in a formal manner for resolution. If the schedule does not permit communication with the customer, the proposal manager must make the decisions and all team members must abide by them.

Another potential crisis may be caused by the discovery that some specification detail is "impossible" to meet. The proposal manager must decide which trade-off will be proposed. For example, if the weight is impossible to meet, offer to trade more strength. If the competitors are no smarter than you, this will not hurt the proposal. Whatever the concern, don't stop the proposal activity because of an "impossible" specification. The customer will buy something. So, offer the next best thing to the "impossible" requirement and avoid the usually fatal error of many "pure" engineers or programmers. Do not play back the requirement to the customer, but tell how the specification will be met.

The trade-offs which have been considered and the reasons for the selected approach mentioned in the introduction should be given detailed consideration in the body of the proposal. One commonly considered trade-off is whether one should offer standard equipment and take exception to the specifications or offer a custom unit that is in strict conformance to the specifications. Discuss the advantages of each alternative in such a trade off and give the reasons for the choice. Also point out that if the customer disagrees and feels that their purposes would be served better by another alternative, you will be glad to discuss it and provide a price. Remember that the first bid should be compliant and alternative bids can show less expensive or better alternatives.

There are several advantages to this discussion of trade-offs. It casts doubt on the competitors who use the alternatives you have discredited. It explains the higher prices the offerer may charge for the choices made. It emphasizes some of the advantages of the system. It lets the customer know that the offerer has done some of their systems engineering for them and they have much to gain by considering the company as a supplier.

A convenient form for discussing trade-offs in the proposal is illustrated in Figure 4.1. First, state the factor being considered, then show the choice of solution on the left and on the right the rejected approach (which may be the competitor's method, or a lower cost method, or whatever the writer wants to shoot down).

THE PROBLEM

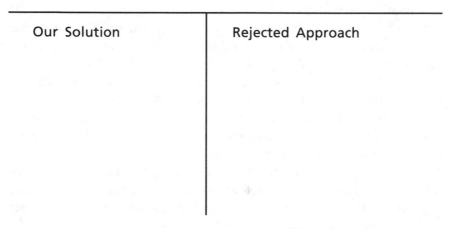

Our Solution	Rejected Approach

Figure 4.1   Perfomance trade-off chart

Technical Capabilities

Start the technical capabilities section by restating the customer's requirement. Although they know what it is, this gives them the assurance that the writers understand it too. In fact, some of the top managers at the customer organization may read the proposal, so make it stand on its own. Show a technical understanding of the customer's problem. Why is this device or this system necessary? Provide more than one solution to the problem and then explain why this one was selected.

Describe the novel aspects of the solution. Are standard products being proposed? Then show why they were chosen. If they are modified to meet this requirement, show how and why they are modified. Use photographs or sketches. If there is reason to be proud of the controls and indicators, *show them in sufficient detail that the customer can read the lettering.* Perhaps the back of the unit should be shown; all those connectors mean flexibility. Avoid over-engineering and unnecessary sophistication. Meet the minimum requirement at a minimum price, then offer alternatives to improve the offering at a higher price.

Make certain that the hardware description is clear. Do likewise for software, making the software section readable for the engineer as well as the programmer.

Unless the customer document is extremely clear, make a thorough statement of work. Otherwise, it can result in unfavorable or arguable contracts. Show quantities, defining each piece clearly. Show all the services which are offered. Make certain that this description is understandable to the intended reader.

Are there deviations from the specification? Show each as a benefit to the customer. Again, be careful not to deviate where it may cause the proposal to be disqualified.

Does the customer doubt that bidders can meet the schedule they have defined? If so, make sure to address this facet of the proposal. What will be done to ensure on-time delivery?

In discussing the maintainability of your equipment, discuss diagnostics and how they are conducted. Show all the features that make it easy to find and repair a problem. If reliability is a factor, define the mean time between failures (MTBF) and explain how it is calculated. Clarify how you will provide spares recommendations on the equipment. After all, spares represent another increment of profit.

Finally, even though the customer didn't ask for it, show some of the expansion features of the equipment. Show the spare input and/or output ports, the space for additional modules, the add-on software. It may gain some points, even though it was not required.

Clarify Your Writing

"Understanding" and "believing" or "accepting" come down to pretty much the same thing. A writer can't compel a reader to accept anything, but can use persuasion to lead readers toward that acceptance.

Logic is not persuasion, although a proposer does need logic as a backup. The proposer must provide the reader with a reason to want to believe the arguments.

The writer loses the chances for persuasion by confusing the reader, so establish and maintain a distinct theme to guide the reader away from possible confusion.

Theme is not enough. The text must be orderly and must progress along specific lines. The ideas must flow smoothly into one another with the connection between them obvious and the progression continuous.

Use the proposal writer's version of Murphy's law: anything that can be misunderstood will be. Write so that the words cannot be misunderstood. Be sure to know what you are talking about, understand the subject thoroughly, and think the matter out completely before attempting to present it to anyone else. Use language carefully. Avoid misleading idioms, because readers may not understand them. Be careful about the connotations of the words and phrases. If at all possible, have a good editor review the manuscript and help to accomplish the clarity.

SPECIFICATION COMPLIANCE AND EXCEPTIONS

If one can't meet the specification, but the company decides to bid the job anyway, show how this noncompliance can benefit the customer. The customer has spent weeks or months preparing the procurement document and certainly has a right to expect the bidders to read it and respond in detail. A separate specification analysis is an absolute necessity. This will list the customer's paragraph numbers and headings and will state the degree of compliance with each item.

Where the proposer complies with a given paragraph, say, "Comply; refer to page 3-6 of this proposal for details." If it complies and the writer has made no reference elsewhere, say "Comply in this manner: _____". If it does not comply, provide a reason the customer will appreciate. Say "Exception is taken to this requirement, in order to keep the overall weight within specified limits. A specification of _____ can be met by the proposed equipment without requiring an increase in weight." In many small procurements, a sensible exception to a customer specification is not a symbol of disgrace, if the exception is for the benefit of the customer and this is explained in detail. Beware of the tendency to make a dogmatic statement in the exception because your competitor may have found a solution to the "impossible" problem (in other words, you lose!). Watch out especially for procurements in which noncompliant bids must be rejected if there is a compliant proposal.

How can one respond to an unattainable or impractical specification? There are several ways to do it without closing the door in the customer's mind.

Case 1: The customer has specified a maximum weight of 40 pounds. One of the competitors can meet this but your unit weights 45 pounds. Respond in such a way as to hurt the competition by highlighting the benefits of your approach.

Case 2: The customer has written a specification based on a competitor's equipment. Your salesman tells you that it was not an expression of favoritism, but merely insurance that someone could be compliant. Instead of listing all the ways in which you deviate from the specification, make a statement about your equipment, with just one reference to the specification.

Case 3: The customer has asked for an impossible operating characteristic. The laws of physics prevent anyone from meeting the specification. This is a difficult position because you either look dumb by taking exception when someone else can meet the specification, or look smart by showing an understanding of the problem when the competition doesn't. Make certain that you look smart and don't make the customer look awfully dumb.

Example 4-2. WACKY Takes No Exception. WACKY's customer had left a significant typographical error in the bid package that made it impossible to meet the specifications for that item. All the other bidders had caught it, and each of them worded the response very carefully to take note of the error. The proposal writers at WACKY were afraid to take any exceptions (or maybe they just failed to read the specification carefully). The proposal was sent in with a cryptic "No exception" to the item.

In cases where the item in question is probably a typo, it is not necessary to say "Exception." The people at WACKY could have said: "On the assumption that _____ intended the output voltage to be 5 volts rather than 50 volts, no exception is taken".

PERSONNEL RESUMES

You are proud of your team members. Share your pride! The resume section gives the opportunity to capitalize on the unique qualifications of personnel. It is a golden opportunity to get a few more points, except the average resume never gets anyone any points. Spend some time developing a resume file. Then select carefully and customize the appropriate resumes for each proposal.

Lay out the resumes in a standard format. It makes it easy to prepare them and, more importantly, easy for the customer to read and reread them.

Start with the person's name and title or area of responsibility for the proposed contract. Proceed to the education section. Remember that this is a sales proposal so don't bother with the details of education that do not enhance the prospects for this sale. Leave out the person's high school, for example, and any nonrelated education. Leave out trade schools and basic military schools unless their content contributes uniquely to this proposed contract (or unless the person has no college degree). List academic honors and membership in any honorary societies that are relevant. Add a publications and honors section if there are such things to add. Omit the whole topic, however, if there is nothing to say. Don't bother to say, "Publications and Honors: None."

In the professional experience category, list the person's present responsibility first, then work back to the beginning of employment in the company. Show previous employments in reverse chronological order, from most recent to earliest. If a person held several jobs in a brief period of time, consider grouping them within a single listing. Earlier employments that do not demonstrate the person's value for the contract being proposed can also be grouped into a single item.

If the person has insufficient background on the present job, expand it by telling a little about the product or system. For example, John was on the design team for an analog multiplexer/encoder but there is not much to say about the part of the job that he worked on, so instead, tell a little about the product itself.

Tell the truth in resumes, but it is not necessary to detail the whole truth. A resume is a sales tool. Tell the things that are salable about each team member and leave other details out.

What about a personal data section? Leave it out! Birth date, name of the spouse, number of children, and hobbies may be interesting, but they are not salable.

Aim for a standard length for all your resumes, that is, one page, or maybe two. A special exception can be made in the case of a senior consultant or other unusually salable person, but don't include more than one or two of these.

The least desirable person to write a resume is the employee being described, as it usually results in either super modesty or pompous words. Instead, assign a good sales-oriented writer to interview the employee, write the resume, and have the last word!

Finally, fine tune the resumes for each proposal. Look carefully at each person's assignment on the proposed contract, then modify the resume as necessary (factually, of course) to highlight his or her capability to fulfill that assignment.

USE OF TABLES AND GRAPHICS

Narrative text often becomes a boring presentation of statistical or performance data. Tables, on the contrary, present such data in a compact and readily understood format that holds the reader's attention. Use them freely.

The proposal can use illustrations in the same way to enhance the message. Select illustrations that convey the essence of your proposal clearly by their captions and contents. Illustrations with action titles can communicate pages of argument.

Each illustration should accomplish one or more of the following objectives:

֍ show how something works or is organized.

ֆ indicate size or appearance.

ֆ prove or highlight a point or benefit to the customer.

ֆ explain a feature.

Avoid analogy or symbolic illustrations and remember that the reader is an evaluator. First impressions may be based on your illustrations. The message should not depend on either the text or illustrations, exclusively. Both text and art should be properly integrated.

Possible Illustrations for an Airport Windshear Monitor System

In a technical proposal regarding equipment to monitor windshear conditions at an airport, the writer must communicate a message to the personnel who must install and operate the equipment, as well as those who will authorize its purchase. Detailed technical illustrations and words will be lost on many of the readers, so the writer must be careful to address a variety of persons in words and pictures that they can understand.

The next three illustrations are suitable for this type of proposal.

The reader needs to see an overview of something to which he can relate, and to see your equipment in that setting. Figure 4.2, an airport configuration illustration, gives you the opportunity to put him in the picture and to identify by name, location, and function the pieces of a system which you propose to provide.

Figure 4.3 shows the point at which wind measurements originate. It is simple and nontechnical, conveying a basic message without confusing any of the readers.

And even though Figure 4.4 shows the output of a computer, it shows it as a simple layout of useful and understandable wind data. The persons who will monitor this information on the completed and installed system must feel that they can understand it. That is the objective of the simple layout.

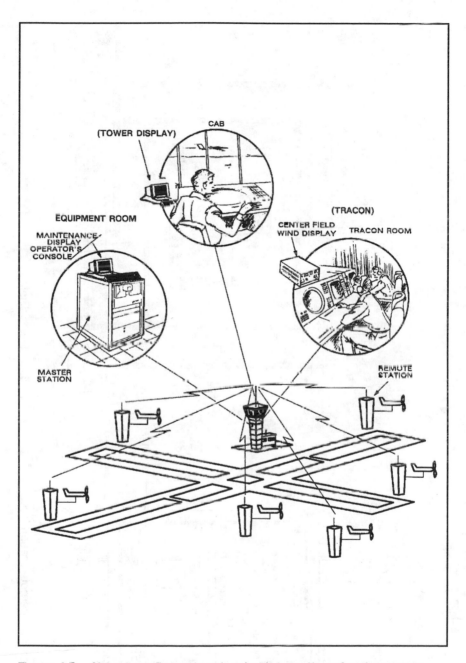

Figure 4.2   Airport configuration showing integration of various parts.

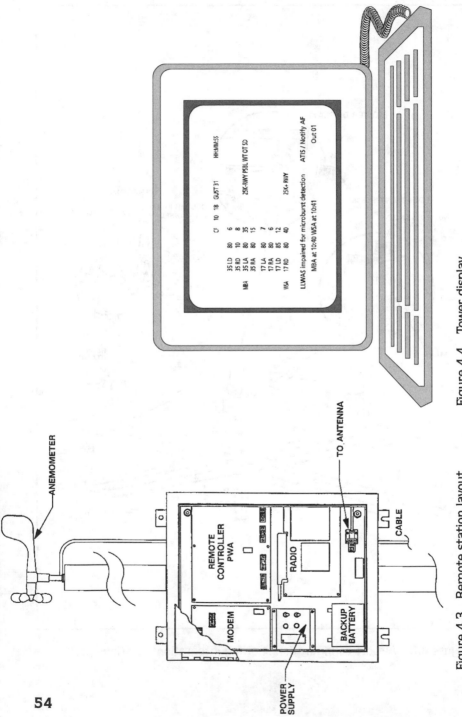

Figure 4.4   Tower display

Figure 4.3   Remote station layout

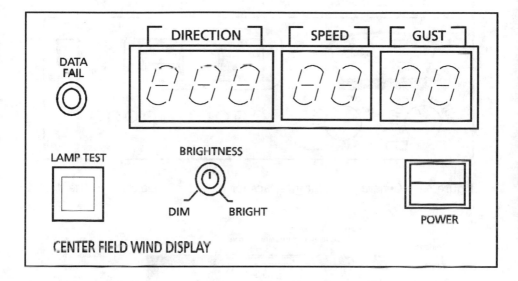

Figure 4.5   Display front panel

An air traffic controller has a very demanding task. The tendency is to resist the addition of new displays in the control room, so the proposal writer must convey the concept that this is a small display containing only the necessary details of wind shear and that those details are incredibly easy to read and interpret. Figure 4.5 above shows the front panel of such a display.

Possible Illustrations for Medical Analysis Equipment

As with other high-technology equipment, the intended user of medical analysis equipment is not likely to be skilled in the technology which is used to design and build the equipment. Thus the proposal writer must not use words and illustrations of the equipment designer, but those of the user. These illustrations relate to a complex medical analysis device, but use simple terms for transfer of information.

Use a simple graphic, like Figure 4.6, to show the operator area on your high-technology device. The reader wants to see how simple (or difficult) it is to control and monitor.

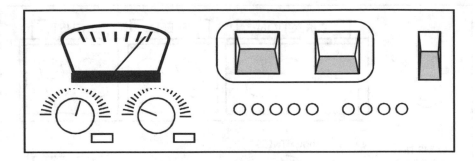

Figure 4.6   General operator interface for a piece of medical equipment

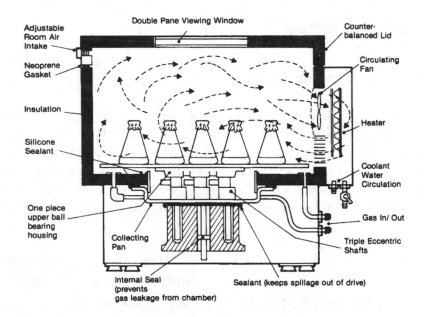

Cold water or low temperature coolant circulates through the heat exchanger for near- or below-ambient temperature. To maintain precise gradient, samples in the incubator chamber are exposed to rapidly circulating air tempered by pulsations of heat on demand. The incubator shaker attains temperatures of approximately 5°C above ambient temperature. Lower temperatures can be achieved in the G25 with built-in refrigeration.

Figure 4.7   Incubator shaker, courtesy New Brunswick Scientific, NJ

Show a typical application with just a few words to illustrate the operating functions, as in Figure 4.7 above.

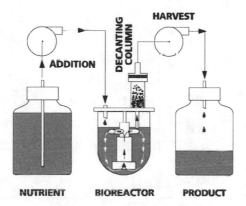

In anchorage dependent perfusion culture, cells are grown on microcarriers in a homogeneous environment in which spent medium is removed and replenished with fresh medium. Microcarriers are gently separated from the harvested product in a decanting column that allows the heavier microcarriers to settle out and return to the culture vessel where cells continue to grow to high density.

Figure 4.8   Cell culture perfusion in the celligen, courtesy New Brunswick Scientific, NJ

Let the reader see how the device accomplishes the tasks that are needed.

Possible Illustrations for
Computer Software Features

By its very nature, computer software is "soft," not concrete, and does not lend itself easily to illustration. Yet to sell complex software in a complex system, the proposal evaluator must be helped to understand the basics of that software. Keeping in mind that such an evaluator may not be able to write a single line of code, yet may control the purse strings of the procurement, the technical illustrations must look relatively simple.

The next two illustrations can be understood by a nonprogrammer.

Figure 4.9 shows the reader the basic functions of a software system; it conveys the message that the system has several meaningful functions and also says that access to each of those functions is quite simple for the nonprogrammer.

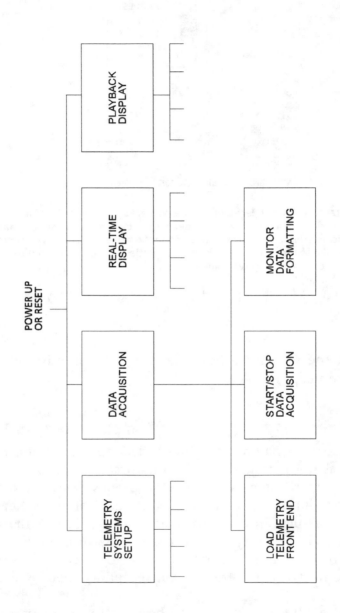

Figure 4.9 Master menu layout

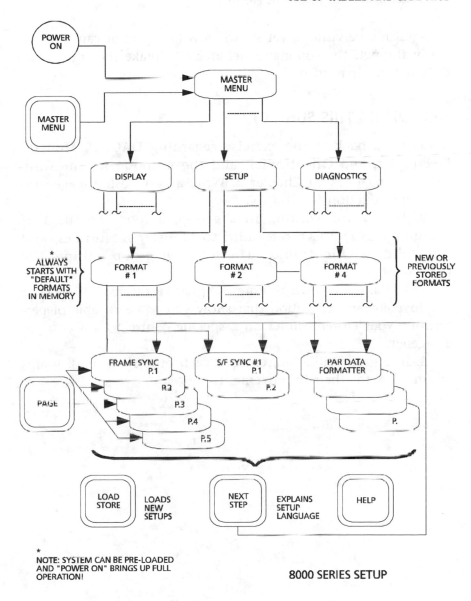

Figure 4.10   Command hierarchy diagram

Since setup of a complex system frightens a nonprogrammer, proposal illustrations must make the task look simple. Figure 4.10 shows all the system functions involved in setup (thereby emphasizing versatility), yet shows how an operator can progress easily through the command heirarchy to make the equipment do whatever is needed.

DEVELOPING THIS SUBJECT

1. Look back at the exercise regarding DACCO, Monster Electronics, and Orientronics, and the proposal for the wind tunnel test facility in Chapter 2. Write a two- to four-page proposal introduction for DACCO.

2. If you are working in a group, assign one third of the group to DACCO, one third to Monster Electronics, and one third to Orientronics. The let each group critique the others' work.

3. Prepare your own resume. Emphasize the aspects of your employment and education that show you to be capable of performing your current duties on a specific contract for a customer agency.

4. In a group situation, critique each others' resumes. Brainstorm ways for their improvement.

5

Your Proposal—
Those Other Parts

L̵ET THE CUSTOMER KNOW
what you will do and how you will do it, to insure that the job is
done well.

DEFINING YOUR PROGRAM
MANAGEMENT APPROACH

Since an offering is for definable items and/or services, there
is a tendency to put off sections such as those on program
management until last and thus to short-change the customer in
these areas. Don't! Consider all of the documentation, such as
operating instructions, maintenance instructions, spares list and
recommendations, interim progress reports, final report, and
drawings. Consider all of the support services such as installation or installation support, training, in-warranty service, and
out-of-warranty service on contract. Plan an approach and spell
it out in sufficient detail to remove any doubts about what the
company will do and who will do it.

Next, tell the customer when the company can deliver each item or service. If one of the deliveries is unbelievably fast, use a few words to explain how the organization can do it (for example, diverting a half-finished power supply from another job). Also, explain any unusual bottlenecks that will pace deliveries. When the age-old question of reliability comes up, the proposal manager must be the conscience of the proposal team. Tell the truth but let him or her discuss any problems with the marketing manager and general manager. Managers must remember that the truthful approach, adequately documented, is more palatable to the customer than an unbelievable and undocumented promise to perform miracles.

CAPABILITY FACTORS

Now that a proposal has told a customer what the bidder is going to do, the customer must be convinced that the company can do the job—that it has "capability." Even if the company has a good reputation with the customer, the proposal still needs to emphasize this. Government regulations require that the evaluators consider only what is written in the propo-sal. A newcomer will want to be very explicit.

The major "capability factors" are listed below. All topics may not be required, but only those that carry maximum conviction to the customer and give him confidence.

Capabilities Topic List

1. *Personnel*
   Education and training
   Experience
   Accomplishments
   Potential
2. *Facilities*
   Space
   Material
   Functional groups
3. *Capabilities: Successful performance of similar contracts (on schedule and within budget)*

History
Performance
Reputation
Activities
Interests
Customer understanding
Resources

A customer wants to know that people are competent, facilities are adequate, and the bidder has experience with similar jobs. The customer wants clear answers and the proposal must provide good evidence.

A cardinal rule for those who prepare capabilities information: *Know your company. Know it in depth.* It's not an easy task, and much digging may be required. But know it you must, for your company and its products are what you are selling.

Facilities

A customer wants to know that a bidder's facilities are adequate to do the job. Answering this with regard to space and material is easy; cite the square footage and the efficiency of your purchasing system. For "functional groups" (e.g., a model shop), the proposal will have to be more explicit. The program management section should be uniquely related to this specific bid set. Do not use boilerplate.

Personnel

Introduce the program manager for this job (or the project engineer, or whoever will be in charge) to the customer. What is his or her background in general (the resume comes at the end of the section)? What is the status of a program manager in the company? If a program manager has the full, open backing of the president or general manager, say so. The customer wants to hear that the program manager has total control of the job, reporting only to the top level of management and functioning as general manager for the specific program.

Introduce the contract administrator. Explain about this person's background and how the contract will be administered.

Tell how the program manager and contract administrator relate to each other in your management structure and how each relates to the customer engineer and buyer, respectively.

Next, introduce the program team, by name and by position, on an organization chart. Show the reporting structure. The proposal may show a conventional line organization with the program manager at the top of the stack, or it may show a matrix, with the program manager at the top of the "dotted line" stack, yet each person reports on a "solid line" basis to a full-time supervisor. Either way is acceptable, provided the project manager is truly in charge and operates with the authority and support of top management.

SYSTEM DESIGN AND SCHEDULE

Define the system design and product design activities as they relate to this job. Show that the appropriate design personnel will be available when the contract is awarded, as well as software personnel.

Does the bidder expect to buy significant parts of the system? Show the relationship with the selected vendor or vendors. How can their deliveries be controlled? Define the subcontract management in terms of technical aspects, schedule, and cost.

If a customer wants formal progress reports, explain who will provide them and describe what they will cover. If no formal reports are needed, offer to provide letter reports monthly. This keeps the communication channels open. If there is a formal design review, what type of report will be generated? How can the company initiate the no cost changes agreed upon during the meeting?

Describe the test plan as it relates to this program. If the customer requires specific approval, who will write the plan and/or procedure, and when? Who compiles the test records? What about software for use in running the acceptance tests?

What steps are involved in getting the equipment ready for shipment? Who will carry them out? How will the equipment be shipped? Who pays for it?

In most cases the RFP has provided the customer's schedule. The end date or period of performance tells how much time is allowed for the entire project. The proposer will usually have at least a few intermediate dates that represent customer-perceived milestones.

It is a mistake simply to acknowledge and promise to adhere to the schedule provided. Instead:

1. Provide evidence of faithful study and planning. Providing your own schedule supports this idea and adds to your credibility.

2. The schedule must be entirely compatible with the bidder's own program, including all its plans, procedures, and perceived tasks and subtasks. When building a schedule be sure not to violate any of the customer's important required deliveries. Do not seize additional months of time arbitrarily and do not destroy credibility by offering a schedule that does not match the proposed program.

3. The customer often provides a description of schedule requirements that is either not truly specific or includes potential problems. This is by far the most important reason for developing a unique schedule of your own.

Talk about installation support at the customer facility. How many people will be required and for how long? What should the customer do to prepare for your installation services?

WRITING YOUR ASSURANCE OF QUALITY

Respond with at least the same number of pages which the customer used to define his quality assurance requirements.

Match Your Customer's Specifications

The quality assurance part of the customer's specification will have been written by a quality assurance specialist. It is a necessary part of most bid packages, and must be addressed appropriately.

A good rule of thumb is to respond with at least the same number of pages the customer used to define the quality requirements. If it is a small procurement with a vanilla-flavored statement that the bidder should have a quality program in accordance with a given specification, a canned response may be suitable. If it calls for special quality checks, however, respond with a detailed description of how the company will conform to those requirements. Don't overlook this opportunity to get a few extra points, with a customer who is obviously willing to pay extra for this service!

As you define the quality program, describe the basis of that program. What methods are used on typical jobs? If this one requires a different approach, what is the plan to incorporate that approach? Explain the interaction between quality assurance and production, engineering, programming, drafting, test, and preparation for shipment. If major vendors or subcontractors are used, how does the company monitor their quality programs? Describe the standards laboratory and measurement equipment.

Explain the Organization and Personnel

Tell the customer about the organization of the quality assurance department and how it fits into the overall company organization. Supply resumes of all quality assurance personnel who are involved in the proposed program.

What is the attitude of top management toward the quality assurance department and program? A brief statement by the general manager might be appropriate. Define the results of the quality program? Does the company actually experience high quality in all phases? Say so! What about reliability? What are the capabilities in this line? Tell about these activities and the goals the group has met.

CORPORATE CAPABILITIES

Describe the aspects of your company you would show the customer if he visited your plant. The description of corporate capabilities can be compiled from "file material" but it should be tailored for each particular job. File material should be updated frequently to show the most recent experience and to highlight any new product area.

Technical Details

Your writeup on corporate capabilities for a high-technology company should emphasize the technical details. What kind of laboratories do the design teams have? What kind of test equipment (in general), support facilities, computer facilities, and library materials (in general), do they have? Move on to the production facilities. What does the company have to facilitate adherence to the schedule? To ensure accuracy? To control the processes? How about test equipment? Is it traceable to the Bureau of Standards? Is the company automated in the production test area?

Personnel

Discuss personnel, starting with engineering. What general comments can be made about education, experience, and tenure with the company? Do the same for programmers and for technicians. Don't forget to mention your technical support groups such as training, technical writing, and any others which are relevant to this program.

List and describe management and their dedication to contracts such as the one on which you are bidding. Did they come up through the ranks? Brag about it! Did they come from the customer area? Or the competition? If one came from a "key" competitor, make a point of it. It may make the customer wonder why he or she left.

Production

What kind of production department do you have? Is it unionized? Highlight it if they are not. What is the turnover? Stress this, if it's low. Go on to a discussion of testing, quality assurance, and field service.

Does the company ever use subcontractors? How does the company manage them? Which one or ones are dealt with most frequently? What about consultants?

Describe the facilities. How many square feet of floor space? Is it a comfortable and safe working environment? What are the main products? When and how did the company start producing them?

The corporate capabilities discussion should tell something about the background of the company. Make it interesting and you'll begin to develop rapport. When did the company start, and where, and how? If any of the founders are still with the company, that shows management dedication. Where did the operating capital come from? Is the company publicly owned? Does it have a parent corporation with deep pockets? If so, it assures the customer that the company will not default when the going gets tough.

TECHNICAL MANUALS, TRAINING, AND SERVICE

Are the technical manuals, training, and field service far better than most? Say so! In a high-technology field, especially, these items take on great importance. Recognize this in the proposal.

Supply resumes for the technical writers. Provide outlines of the typical hardware manual, the typical software manual, and the approximate size of each for a specified part of the system. Explain the philosophy of manual development and the quality of the end product.

Provide resumes for the training staff. Explain the general qualifications of the instructors (even if this is an *ad hoc* assignment in your company). What type of training facilities do you have? Show a tentative training outline and provide examples of the types of visual aids and handouts used.

Field service must get the same attention, starting with resumes for the personnel. Who will the company assign to this specific job after it is delivered? When do they assume responsibility? Detail in-warranty service and show options on after-warranty coverage. How does the availability of spares back up the service team? Will there be computer terminals on site with access to the scientific computer(s) in the plant?

USING AN APPENDIX

A typical proposal writer tends to overload the heart of a proposal with technical details which distract readers. To avoid becoming so hopelessly involved in the technical details of a specific section, put technical details in the appendix to the proposal, where they can be referred to but do not detract from the continuity of the proposal's theme. Appendices are useful for materials such as:

1. Back-up information, including long tables, explanations of mathematical concepts, drawings, and lists.

2. Supporting data and discussions for any section that exceeds a reasonable number of pages.

3. Detailed but germane and cohesive information that can be studied later by the reader or by specialists in whose domain the subject matter falls, so as not to disturb the proposal's thematic flow.

4. Data of a temporary nature, such as performance requirements for an early test model of a configuration end item.

5. More complete description of alternative technical proposals.

DEVELOPING THIS SUBJECT

1. Prepare a corporate capabilities section to describe specific departmental functions in the company. In a group situation, critique each others' write-ups. Pool your criticisms to improve the individual write-ups.

2. You are a DACCO engineer, responding to a customer's request for proposal in which the specification is written based on a Monster Electronics product. You feel that a DACCO standard product will serve the intended purpose as well as the competitor's product, but ten different items in the specification define details with which your product is noncompliant. Write a specification analysis section to respond to the requirement. Use your imagination to describe two or three items that the customer may find more important than those defined. Present a blanket noncompliance. Make it appear to be the customer's advantage to choose your standard product.

CHAPTER

6

Cost Estimating and Pricing Strategy

$$C$$*OST ERRORS CAN BE AVOIDED*
by planning, scheduling, and following through.

SCHEDULING AND ORGANIZATION

One of the most difficult tasks in preparing a proposal is estimating the cost. It is often postponed until the last day (or night) and allotted only a brief portion of the time it deserves. Cost preparation can be made less painful by proper organization at the outset.

At the kickoff meeting, or as soon afterward as practical, the proposal manager and the cost accountant should compile the Work Breakdown Structure (WBS), a tentative list of the cost elements and quantities for hardware and software services. If any special breakdown is required by the customer, this tentative list should be in the customer-specified format. All departments who have inputs to the cost estimate should prepare their estimates in the specified format. Further breakdowns may be supplied, if necessary, but this basic layout must not be ignored.

Typical errors in cost proposals are legion: the cost accountant can misunderstand the quantities required; panic due to last-minute preparations can make logical thought impossible; obvious mistakes can remain unrecognized when hours and material are converted into dollars; separate organizations within the company can duplicate the cost estimates for identical tasks; and significant cost elements can be overlooked in their entirety. All of these can be avoided by planning the cost estimates, scheduling them, and then following the plan and schedule. Use the work breakdown structure (WBS) described in Chapter 3, and key all the cost items to WBS numbers.

PRICE CHECK LIST

1. Is this the lowest possible price, considering long-range potential versus immediate return and probable competitive price range?

2. Have all "make or buy" aspects been considered?

3. Have subcontractors and vendors submitted their lowest realistic prices?

4. Have hours, space, facility, and other cost factors not been overestimated?

5. Has consideration been given to the dollar value placed on the project by the customer and the funds available for it?

Difficult as it may be, cost estimates must sometimes be completed very early in the proposal schedule so that trade-off studies can be made. The team can't afford a last minute surprise but early surprises can be handled comfortably. Early cost estimating has changed many proposals from losers to winners by sending the teams back to the drafting board for new systems approaches.

General Evaluation of Job Costs

1. Which direct charges to programs and projects for specific customers contribute to meeting the specifications of the contract? Which do not?

2. Which of the various indirect charges contribute to overhead or general and administrative expenses? Which could you justify to a customer as contributing to your viability as a supplier?

3. What levels of personnel do you bid and assign to jobs? Can some of the tasks be done properly by more junior personnel?

4. How significant is the "fear factor" in your bids? Do you bid high enough to guarantee meeting the budget, or do you bid so that the average job meets the budget?

5. Do you bid "creeping elegance" into your jobs because you enjoy doing more sophisticated engineering and providing a more attractive system? Do your competitors do the same (or do you just hope they do)?

6. What do you ship in each system that gives your customer a pleasant surprise? Should you eliminate these extras? Should you document them very clearly in writing, so that your sales engineer can ask your customer to specify and demand them of everyone? This may apply to hardware, software, technical manuals, training, field service, or some other area.

7. Is your software structured so that much of it is usable on future systems? This is the way to make money on software. The programmer's tendency is to continue to improve a package and to resist using it a second time without such improvements. Can this luxury be justified in a competitive world?

Example 6-1. One of the upper level managers at WACKY had a strict "no cost overruns" policy for his project people. Any project engineer who overran a budget was criticized privately and publicly for this sin, not just once, but continually. This policy caused a sequence of events:

1. People began to agonize over their jobs, working extra hours to stay out of trouble.

2. It became impossible for some projects to stay in the clear, even when the staff worked evenings and weekends. The next solution was to falsify time cards and material charges, in the hope of fooling the system and preventing criticism.

3. Some of the best people fell victim to these practices and began to ask for transfers and even began to look for transfer to lower pressure environment in another company.

4. Those who stayed found their jobs more and more difficult and began to react by padding their cost estimates on proposals for new jobs.

Eventually, business dropped off because the bids were too high. Burden rates went up because of the reduced business. People began to panic, and more good employees left.

All this could have been avoided by an adherence to a simple, sensible rule: "Perform within budget on the average job. If you overrun one and under-run the next, you will not be criticized." In the words of a vendor at the stadium, "Make up on the peanuts what you lose on the popcorn."

Remember: Help people to do their jobs efficiently, and praise their good results. Shoot for the average job to be within the budget, not every task and every subtask. The people and the company will profit by it.

7

Putting It All Together

T*HE ENTIRE PROPOSAL SHOULD*
read as if it had been written by one person.

PROPOSAL—FINAL READING

As the short, concise proposal is completed, it must be reviewed by one knowledgeable person who has read the customer's requirements in detail. This review is not intended as a general proofreading operation but specifically for catching any major errors or omissions. The review should verify all references to illustrations, to the appendix, or to other sections within the proposal. It should look for conflicting statements. It should verify that the cost proposal is in conformance with the statement of work in the technical proposal. When the reviewer has finished, the entire proposal should read as if it were prepared by one person.

The large and complex proposal deserves a more thorough review. A technique used in many companies is evaluation and scoring by a "Red Team," an *ad hoc* group of knowledgeable persons who are not members of the proposal preparation team. The Red Team members evaluate the draft proposal in its entirety against the customer's RFP, and evaluate it without regard for possible bruised egos. They should read it the way the customer will. Many proposals have been changed from losers to winners by these 3 or 4 days of detailed evaluation and the resulting cleanup.

PRODUCTION DETAILS

Plans for final printing of the proposal should be announced well in advance of its completion so the printers can make significant preparation before the material begins to arrive. As each section is completed, it should proceed to reproduction. It will be far easier to change two or three sheets to correct errors than to hold up the entire job for an extra day.

The proposal should be printed on a good quality paper, using an easily readable printing technique. The schedule may require some compromises with the publications group, but their suggestions will be of great value as the manager selects these compromises.

If most or all outlines follow the same sequence, buy pre-printed section dividers and use them on all proposals. They don't cost much and they add significantly to the end product's sophistication.

Sequence of Contents

The cover should be neat but not gaudy. Its design reflects the advertising image of the company. Its simple purpose is to announce the contents, identify the sender, and protect the document. Modesty is the keynote. The value lies in the content of the proposal, not its trappings.

The title page repeats the cover title, refers to the bid request, identifies the sender and includes a complete address.

The abstract, or executive summary, is the keystone of the proposal. It summarizes the content, but does so selectively and with hard facts that back up the sales message. Imagine that the writer is answering the customer's comment: "I don't see anything about your system that's better than any other system. Tell me in 300 words or less what's so good about it." Or, as

Mark Twain's Jim Smiley put it, "I don't see no p'ints about that frog that's any better'n any other frog." Answer that comment, and you'll have a good summary.

The summary serves other valuable purposes besides its brief appearance in front of the proposal. A copy can be attached to the letter of transmittal as valuable supplementary information. Also, copies can be circulated within the company as general information.

The table of contents gives customers the opportunity to read, reread, and then read the proposal once more. Don't forget, this is one of several proposals on their desks. If they see something in a competitor's offering that looks good they will want to check each again and again to see how various subjects were addressed.

Put every important topic in the table of contents. That doesn't mean that one should go to the fourth or fifth level of subparagraph. Use whatever level is appropriate to make it easy for the customers to read and reference. Include a list of illustrations, tables, and other enhancements if they are helpful.

COVER LETTER

Even though the cover letter is likely to become detached from the proposal at the customer's procurement office, consider it a sales tool. Some companies go so far as to say that they win or lose based on the quality of their cover letter! It should be prepared well in advance of the deadline for mailing.

Use the first part of the cover letter to emphasize the strong points that have been defined in the abstract. State them strongly for the benefit of all who will read it.

The cover letter should be signed by (or at least "for") a top management person who can "make it happen." It should include a commitment that the corporate management desires to get the contract, and the designated project manager will be delegated the authority needed to be successful.

The cover letter should indicate the basis of the proposal (unsolicited, or solicited by Request for Proposal Number ____), and the price validity period. It must make any specific representations that may be required by the customer. It lets the customer know who to contact for additional details or for clarification. It should state your offer to give a verbal presentation at the customer's location. Last, but not least, ask for the order.

TIME AND MONEY SAVERS
TO USE IN PROPOSAL WORK

Meetings have been held, tasks assigned, and the proposal course has been charted through calm waters. Now all hands set to work to produce a manuscript that can be sent to the printer.

How does the manuscript take shape? Around the outline, of course. A good technique is to use an ordinary three-ring notebook, with separators keyed to the major topics of the outline. As input is generated, it is slotted into the notebook. Every illustration and exhibit is represented either by the actual item or by a "voucher" page. This system ensures that nothing is left out. Also, this complete manuscript serves as a guide for assembling of the printed copies.

Using Stock Material

Certain proposal inputs remain unchanged from one proposal to the next. Save time and money by printing such material in quantity and storing it for future proposals. These include corporate brochures, professional papers and articles, and any relevant advertising and sales materials.

Stock material should not be looked on as "filler," because it plays a vital role in the proposal. It must be selected with care and updated frequently to reflect the latest and best information. Most important is that it be tailored to the requirements of the bid request for each proposal. If this is not done, it deserves the name "boilerplate" and its weight can sink the proposal.

Proposals are built upon each other. To take maximum advantage of this fact, extract typical sections from past proposals, such as summaries and introductions, and store them in a notebook with the source document number marked on each page for ready reference. This material is valuable as a writing guide but cannot be used unchanged from proposal to proposal. Fortunately, the word processor makes changes and additions easy.

Make up a checklist of resumes. The proposal manager checks off the names to be represented in the proposal and the resumes can be collected, modified, and held for assembly.

Organizing the Sections

Number pages by section rather than sequentially from first to last. For example, the first page in Section 4 is numbered "4-1". Some sections that are easier to prepare can then be typed and ready for the printer even though they don't occur at the beginning of the proposal. They remain undisturbed by the more problematic sections. If last-minute changes are made and pages added, only one section need be affected. This practice applies to illustrations and tables as well. Standardize formats and procedures so that typing and other processes become automatic.

Use all possible shortcuts in preparing artwork, such as tapes, preprinted transparent overlays, shading sheets, and other aids. Print or bind only those copies needed for the customer, and hold internal distribution for later.

Appoint a "proposal specialist" (a technical writer is a good choice) who is steeped in the art and assumes responsibility for all the mechanical aspects of proposal preparation.

Help Your Writers

The product for sale is technology. The proposal will be no better or worse than the abilities of the contributing engineers and programmers. A direct route to better proposals is to give the engineers and programmers maximum assistance:

1. Set up an efficient system for proposal preparation. This frees the engineer to concentrate on one task and one deadline.

2. Give memory joggers in the form of themes and topic lists, as described in previous chapters.

3. The team member often needs information but has no time to dig for it. The company information retrieval system can be invaluable.

4. For some proposals it may be helpful to set up a quiet room for the proposal team, away from daily distractions, and

give writers a chance to communicate, plan, and write with minimum of wasted time and effort.

5. The product for sale is technology, and the winning idea may come from an engineer who kept abreast of the latest developments.

Aspects of Style

A good technical proposal cannot "sell" equipment, but a poor one can "unsell" it. Everyone knows this, yet proposal after proposal is generated with the same basic mistakes in the technical section. For example, the opening is so lengthy and dry that the customer never gets to the meat of the subject. Or, the proposal fails to address the real problems. Or, illustrations are used improperly. Or, too much detail is included in the body of the proposal instead of the Appendix. Or, the proposal is a discussion of the impossibilities of the problem instead of a positive discussion of its solution.

For maximum impact, a technical proposal must use active statements. Describe the equipment in the present tense: "The phase lock loop is a single integrated circuit," instead of "The phase lock loop will be an integrated circuit." The customer knows that you have not actually built this system, but they are influenced positively by a positive approach.

Emphasize why the company made certain choices. To state the use of a given circuitry is not illuminating. Instead, provide a list of several types of circuits which could have been used, the advantages and disadvantages (to the customer) of each, and the reason for the final choice. After all, the competitor may plan to save a few dollars by using an inferior type of logic and, surely, you don't want the customer to fall for that one!

Prepare the technical proposal with short, snappy paragraphs that are easy to read. Don't forget that the customer will be evaluating a stack of proposals. Make this the one that "wants to be read." Use the same techniques the editor of a daily newspaper uses to hold the reader's attention. You aren't likely to read the paper from beginning to end. If a paragraph is dull, you skip to the beginning of the next paragraph. If an article is dull, you look at the headlines until you find the next interesting one. Your eyes are almost certain to lock onto the articles that are illustrated or that contain tabular information. A well-planned layout will hold the reader's attention, and recapture it frequently.

Since the proposal will be read, and reread, and then read again, make it easy for the customer by indexing the sections by subject. Illustrations should be indexed separately.

Using the Word Processor

If the word processor technique wasn't invented by a proposal writer, it should have been! The word processor is the proposal writer's best friend. Use it properly. Every proposal group should have word processing capability, with more than one terminal, a letter quality printer, some experienced operators, and a helpful and innovative supervisor, giving top priority to this project.

Your first proposal on the word processor will go most smoothly. The big frustration caused by last minute corrections becomes a minor problem on the computer. Just change the word, paragraph, or page, and that's it. But if that's good, just wait for your 20th, 50th, or 100th proposal on the computer! You begin to reuse material on the second proposal and begin to build up a resource file from which to "cut and paste."

To get maximum benefit from the word processor, talk to the manager of the group. The way to get the most from this group is to find out what will make the operators' jobs easier. How do they file their disks? By proposal number? Writer? Date? Mark all resources in the same way, so the disks can be accessed easily and quickly. When putting an excerpt from a proposal in a looseleaf resource file, mark it the same way the disk is marked. Always save that identification for the benefit of the word processor group.

How should one mark corrections? A black pencil or pen is not the best way. Find out what works. Use proofreading marks that are most easily understood. It's not hard to learn and it saves minutes (or hours) as that confounded deadline comes nearer.

If Section 1 of the proposal is the last one to be written, schedule it out with the word processor operators. Give them the page budget, and number the pages of each section beginning with "-1." This way, Section 1, page 1-1 can be the last and there is no panic.

Use the word processor and get the most from it. Word processors were made for proposals.

CHECKLISTS

The proposal should be checked carefully before printing.

A. Technical

This section should provide an analysis of the problem, a discussion of the operational environment, and an accurate and clear technical description of the proposed system and/or hardware, including drawings or sketches of the proposed configuration. The following should be considered:

1. Is there a clear concise statement of the technical requirements fulfilled by the proposal?

2. Is the technical problem, as seen by the customer, clearly delineated?

3. Does the proposal show a convincing depth of understanding of the problem?

4. Is there a brief discussion of alternative solutions explored and rejected and the reasons for rejection?

5. Is there a discussion of any new technical approaches to be explored and why this approach can be expected to yield the desired results?

6. Have unrealistic and unreasonable performance requirements that drive the price up been identified and alternatives suggested?

7. In the event of alternative approaches, is the detailed logic for these recommendations given, especially in terms of benefits such as enhanced performance, lower costs, greater produceability, earlier delivery, and simpler maintenance?

8. If certain problem objectives are incompatible with other problem goals (for example, simplicity versus accuracy), does the proposal show that, all factors considered, the optimum solution has been attained?

9. Have the more difficult areas been identified and detail provided to show how performance requirements never before achieved will be met?

10. Is there description of novel ideas or technical approaches?

11. Is there a statement of major technical problems that must be solved, with an indication of the amount of effort budgeted to each?

12. Does the proposal show the proposed solution's relation to the broader overall system with which it will operate?

13. Is there a description of the hardware to be furnished?

14. Is there a realistic estimate of performance?

15. Does the proposal state where it deviates from specifications and by how much? Why does this benefit the customer?

16. Does the proposal show that proper consideration has been given to serviceability and ease of maintenance?

17. Is an estimate furnished of maintenance procedures and schedule, showing to what extent special test or support equipment will be required?

18. Does the approach consider logistics and long-range maintenance?

19. If new components must be developed, does the proposal explain why existing ones cannot be used and how new ones can be developed on time and within the customer's budget?

20. Are unique or unusual component reliability requirements (exceeding those obtainable from conventional components) described and justified?

21. Does the basic proposal meet the minimum requirements at the minimum price?

22. Does the proposal explain why alternatives offered at a higher initial price are cost-effective over the product's life cycle?

Ability

This section should clearly demonstrate the overall technical competence of your company to complete successfully the specific project.

1. Is there convincing assurance of specific technical competence for this project?

2. Are there specific examples of similar projects your company has successfully completed?

3. Do the resumes relate specific experience of personnel to the specific needs of this project? Has extraneous biographical information been eliminated?

4. Is the availability of specific people clearly detailed in terms of hours for both full time and part time staff?

5. Since the customer knows that the same personnel are used for different proposals, does the proposal show a depth of qualified personnel?

6. Are areas of technical weakness identified, and does the proposal show how these will be overcome, for example, by subcontractors or consultants?

7. Does the proposal clearly indicate that there is adequate technical space and facilities, both general and program specific, to perform work efficiently and on schedule?

8. Does the proposal outline the availability of facilities necessary for the specific project, for research, development, production, and testing?

9. Does it clearly spell out any special technical facilities required by the project (such as dust-free laboratories, temperature-controlled rooms, data processing equipment, and special laboratory equipment)?

10. Is it clearly indicated that all facilities will be available when required for this project?

11. Where customer-furnished equipment is required, are these needs clearly identified and justified?

12. Where relationships with subcontractors are proposed, is specific evidence given of the subcontractor's commitment to make technical people and facilities available when required?

Scheduling

Delivery is most important. The proposal must not only state that the delivery schedule will be met but also show how it will be met.

1. Does the proposal provide convincing assurance that the customer's delivery dates will be met or bettered?

2. Is sufficient detail given regarding master scheduling, programming, follow-up, and other similar functions to reinforce the ongoing assurance?

3. If subcontractors and major suppliers are involved, are sufficient safeguards built into the proposed scheduling system to ensure subschedule compliance with the master program?

Management

The proposal should show your method of management. It should elaborate on organization, personnel, and manpower controls. It must demonstrate that you have an understanding of the external organization relations with the customer or prime contractor and with subcontractors who will be needed to finish the project. It must outline the overall management concepts employed and the specific type of management to be provided for the proposal project.

1. Does the proposal clearly demonstrate an understanding of the customer's concern with the management of this project?

2. Are details provided on corporate experience, facilities, and personnel?

3. Does the proposal demonstrate that top level management will continue a high level of interest and assume responsibility for successful accomplishment of the program?

4. Is evidence given of management's understanding of how the specific project fits into the customer's overall needs?

5. Does the proposal show capabilities of the management to handle a project of the size contemplated?

6. Is evidence given that top level management has full control of its organization? And that the program manager has full control over this program?

7. Does the proposal show how your interest in this specific project ties in with the company's long range plans as well as with past experience?

8. Does the proposal show the position of the program manager or group in the overall company organization and the limits of authority and responsibility?

9. If no special group is to be formed, does the proposal show the method of operation within the overall company structure?

10. If organizational charts are presented, do they clearly show how the project management will operate effectively on a day to day basis?

11. Is information furnished on the type, frequency, and effectiveness of management controls and methods for corrective action?

12. Is a total staffing plan furnished, as well as individual plans for engineering, manufacturing, and quality control?

13. Is information provided to show how the present project will coordinate with current and future business?

14. Is a make or buy program provided?

15. Does your evidence support the selection of subcontractors for their technical and manufacturing capabilities as well as their management philosophy and talent?

16. If the proposal involves systems management, does it show how the subcontractor's management will be integrated into the program?

17. Are subcontractor organization charts furnished for subcontractors, showing clearly their relationship to the prime contractor and to other subcontractors?

18. If subcontractors are to be used for a major part or subsystem, is a copy of their proposal furnished or, alternatively, evidence to show that their proposal has been properly developed and evaluated? Have you defined clearly your methodology for their control?

Quality Assurance

The term "quality assurance" covers all the actions necessary to determine adequately that product requirements are met. Quality control is the system and management function by which the contractor ascertains and controls the quality of supplies or services. Reliability is the ability of an item to function without failure. The proposal should carefully delineate the company's programs to address these customer concerns.

1. Does the proposal describe your quality control plan, including organization, policies, facilities, operational system, technical capabilities, and records system?

2. Is it clear that the customer requirements will be met by your quality control system, organization, concept, and approach?

3. Are deviations from customer requirements shown to be covered by equivalent or improved techniques in your company?

4. Will the customer's reliability requirements be achieved by your concept and approach, including a specific program for meeting or surpassing these requirements?

5. Do you show clearly how the reliability organization and project responsibility fit into the proposed program?

6. Are reliability monitoring points (breadboard, experimental, development, service test, prototype, and production) clearly delineated so that customer surveillance may be exercised effectively?

7. Does the proposal show an understanding of reliability prediction techniques and spell out in detail how predicted goals will be met?

Price

The price should not be arrived at by adding the raw estimates of the various company departments. It should be based on the lowest price that will make the business acceptable to the company in light of available alternatives, adjusted to take into account the long run benefits and drawbacks.

1. Is this the lowest possible price considering long range potential versus immediate return and the probable competitive price range?

2. Have all "make or buy" aspects been considered?

3. Is there complete satisfaction that subcontractors and vendors have submitted their lowest realistic cost estimates?

4. Are you certain that hours, space, facility, and other cost factors have not been overestimated?

5. Has consideration been given to the dollar value placed on the project by the customer and the funds available for it?

6. Have the cost estimates been "scrubbed" thoroughly?

Field Support

The field support that will be or can be provided to place the item in service and maintain it in operation must be fully described.

1. Does the proposal adequately cover all aspects of support required for the stated program? The following items should be considered: maintenance, engineering, technical training, technical data, installation support, depot support and implementation, sustaining engineering and product improvement, field representation, provisioning of unit spares and maintenance and operating parts, test, and other ground support equipment.

2. Does the proposal highlight the magnitude and scope of your field service and support capability?

3. Are recommended support aspects described and delineated?

4. Does the proposal provide specific examples of your company's accomplishments in the field service support area?

5. Does the proposal describe the type of support that will be required from the customer?

Manufacturing

The proposal should show your company's competence to manufacture the item. Some information in this field is important even in proposals that may not involve any quantity production, since the buyer must usually give consideration to and plan for future production quantities.

1. Does the proposal describe your manufacturing organization's responsibilities, tool policy and plan, fabrication and assembly plan, quality assurance, and configuration and manufacturing controls?

2. Does the proposal explain the system and procedures used for schedule planning and operational controls?

3. Does the proposal provide convincing assurance of specific manufacturing competence in terms of this project? Do the biographical data relate the specific experience of the manufacturing personnel to the specific work areas of this project?

4. Does the proposal give specific examples of similar projects your company has successfully completed?

5. Does the proposal clearly indicate the varying availabilities of these manufacturing personnel to the project? If subcontractors and/or consultants are involved, does the proposal provide assurance of their availability?

6. Does the proposal clearly indicate that you have adequate manufacturing space and facilities, both general and special, to perform the work efficiently and on schedule?

7. Are the specialized equipment and processes required for the project given sufficient prominence in the proposal through photographs and descriptive information?

8. Does the proposal clearly delineate the work flow paths from the time the engineering plans are released to the time that items are shipped?

9. Does the proposal show evidence of an effective manufacturing control system?

10. Does the proposal indicate a clearly defined procedure under which you can move quickly to meet any emergency with a minimum of program disruption?

11. Does the proposal call attention to the high standards of the product test procedures used?

12. Does the proposal specifically state that all required facilities are available for the project at this time?

13. Does the proposal provide evidence that the most advanced methods are used in manufacturing and manufacturing support areas?

Editing and Format

While no arithmetic rating is assigned to editorial caliber and format of the proposal, their importance cannot be overemphasized. The information must be presented in a logical, pleasing manner that will give the required emphasis. The proposal is the point of sale and should be prepared and presented to the customer with that in mind.

1. Has the proposal been checked for clarity, logic of presentation, consistency, completeness, accuracy, and emphasis?

2. Are the sentences generally simple in structure?

3. Does the proposal follow the organization of the subject matter in the request for proposal, if such is specified? Otherwise, is there a clear organizational pattern or outline?

4. Has the editing removed all unessential, trivial, and repetitive material?

5. Does the proposal follow the basic elements of good writing?

6. Are abbreviations, acronyms, symbols, and mnemonics clearly explained when used?

7. Is the proposal easy to read? Does it have short, logical paragraphs, frequent headings, and dividers for major sections? Is the use of abbreviations confined to standard words which are readily understood? Would an appendix of frequently used acronyms be appropriate?

8. Are modifiers close to the elements they are meant to modify? Are strings of modifiers avoided?

9. Have consistent page and figure numbering been used?

10. Is the table of contents consistent with the size and complexity of the proposal? Is it easy and inviting to read?

11. Have nontechnical synopses of the various sections been provided for the guidance of nontechnical evaluators?

12. Does the proposal quote references rather than cite them?

13. Do the illustrations contribute to the "story line" of the proposal? Do they simplify its readability and are they functional? Are they easy to understand? Are they effectively located in the text?

14. Do tables meet these same criteria?

15. Is the completed proposal an attractive sales package?

Cover Letter

1. Is it addressed to the contracting officer, in content as well as in form?

2. Does it stress the items a contracting officer wants to see?

3. Does it stipulate that the signer is authorized to make an offer?

4. Does it specify the validity period?

5. Does it invite questions? Does it offer an in-person presentation?

6. Does it ask for the contract?

DEVELOPING THIS SUBJECT

1. Look back to the exercise in Chapter 2 involving DACCO, Monster Electronics, and Orientronics, and the proposal for the wind tunnel test facility. From what you are told, and from your imagination, compose an abstract for your proposal. In a group situation, assign one team to DACCO, one to Monster Electronics, and one to Orientronics. Compile your abstracts and then look at each other's

2. Compose a cover letter for the windtunnel facility. Use your imagination to fill in the details.

8

After the Proposal Is Mailed—The Follow-Up

D*ON'T STOP WORKING when you mail the proposal. There's more work to be done.*

AFTER THE SMOKE CLEARS

As soon as the smoke has cleared after the proposal is delivered, the marketing department should begin a follow up with the customer. The exact method of follow-up depends on the rules of the customer's procurement department. One cannot afford to press a case any more (or any less) than the rules will permit.

After all efforts are expended (and you have won the job) have a brief meeting to analyze the reasons. Make a record of them. What were the weakest points? How can the company make sure that its next proposal will be a winner also?

If (perish the thought!) this one is a loser, hold a postmortem on the loss. Try to find out from the customer why this one was lost (they will usually tell you). Plan the corrective action so that the next proposal is a winner!

FORMAL PRESENTATION

Many customers make it a practice to offer the top company or companies the opportunity for a technical presentation a few weeks after the proposals are submitted. Don't pass up this opportunity! It is a good way to underscore the strong points which were brought out in the proposal.

Ask how much time is allocated for the presentation, and verify that slide or overhead projection capability is available. Then use the best resources to prepare for this important session with the customer.

First, this formal presentation is not an invitation to parade all the proposal team in front of the customer with each person taking five minutes. The technical and program management parts should be a presentation by the one or two (no more) people who are good at making presentations. Select those people carefully, give them plenty of time to prepare, give them the full resources of the proposal team, clear the way for them to get the top graphics talent, and your company will be able to score extra points with the evaluators. Zero in on those strong points and build them into this presentation as a smoothly flowing discussion. Don't restate the proposal. The customer has read it already! Review the strong points in a different way. Use graphics and color as much as possible. Refer to Chapter 12 on technical papers for reminders on how to prepare graphics.

The technical presentation team should consist of the one or two people who will make the formal presentation, plus one or more specialists who will be introduced and may be needed for answers to detailed questions. Add to this a top level manager who will express the company's commitment to the program and the sales person (closest to this customer group) who will introduce the delegation. Leave everyone else at home to mind the store. The first choice of speaker for the main presentation is the program manager if he or she is good at public speaking. The second choice is the proposal manager who should be identified as an active participant on the job.

When fielding questions from the customer on the proposal, do not feel obligated to answer on the spot (unless you had ad-

ANALYSIS

Customer: _____
_____
_____

Competitors:
1. Most formidable _____
2. Next most formidable _____
3. Next most formidable _____
4. Next most formidable _____
5. Next most formidable _____

Brief description of the product(s) or system(s):
_____
_____
_____

Quality of our rapport with customer
_____ (0-10)

Quality of our past dealings with customer
_____ (0-10)

Quality of our proposal _____ (0-10)

Prices (if known):
1. Ours $ _____
2. _____ $ _____
3. _____ $ _____
4. _____ $ _____
5. _____ $ _____

Our significant positive features:
_____
_____

Our significant negative factors:
_____
_____

Other observations:
_____
_____

Result of customer evaluation:
Best technically: _____
Most attractive price: _____

Winner:
Biggest reason: _____

Improvements we should make as a result of this analysis:
_____
_____

vance copies for study). It is better to postpone an answer for a specific length of time than to give an off-the-cuff response that may damage the cause.

POST-AWARD ANALYSIS

A clear and simple way to carry out post-award analysis is to complete a chart similar to the one on page 95. Add it to the proposal writing resource file to help in the preparation of the next (winning!) proposal.

DEVELOPING THIS SUBJECT

Example 8-1. WACKY had submitted a proposal to a government agency for a large system, one of the largest in WACKY'S history. The opportunity for a formal presentation came, and the division manager decided that this was a golden opportunity to present the entire proposal team to the customer. Each team member was given an assignment for a 5 minute presentation. All were told to travel together and to spend the afternoon before the presentation in a large motel room "rehearsing."

The day of the rehearsal was a shambles. Some members arrived too late, some arrived too early (and spent the extra time in the bar), and most felt uncomfortable in a rehearsal atmosphere.

The presentation was predictable. Some of the participants had given very little thought to their contributions. Others had given no consideration to the time constraints and talked far too long. The last ones on the program were rushed and had to make their abbreviated comments off the cuff. The result was a loss of points with the evaluation team and loss of the contract.

Remember: Plan properly and gain points rather than lose them with the oral presentation.

9

Case History of a Winning Proposal

T HIS CHAPTER SHOWS HOW to put the steps discussed previously together to formulate a winning proposal. They use DACCO as an example. The customer is the Interplanetary Rocket Research Agency of the U.S. Postal Service. This company and customers are hypothetical but the case history is a composite of actual steps by real companies with real customers which resulted in winning proposals.

The reader will note that the activity starts long before the bid set is received, and continues after the contract is awarded. Each step in this lengthy process is important. Insufficient attention to details along the way could have negated all efforts to bring in the eventual contract.

ADVANCE WORK ON THIS OPPORTUNITY

The Interplanetary Rocket Research Agency (IRRA) is located in Jupiter, Texas. DACCO has a sales engineering office 200 miles away in Dallas. The engineer in that office has made calls on IRRA since the agency was founded in 1978, leaving technical literature on DACCO and collecting whatever information was available on the character and organization of the agency. He felt frustrated at times, but the occasional small orders and notice of long range plans were enough to bring him back every few weeks. His sales manager at DACCO encouraged him to cultivate this customer, and to establish a technical rapport "just in case."

Our sales engineer's best opportunities to build rapport came with those small orders. He acted as if each was a million dollar sale. He showered IRRA engineers and procurement agents with attention as he expedited each order and followed up to verify satisfaction or to provide any necessary assistance with problems that developed. As a result, DACCO became known to the customer as a technically capable and personally responsive company.

When IRRA began to develop plans for a Rocket Test Stand, the customer engineers looked at available "building blocks" from which a potential system supplier could develop the system. The DACCO equipment already on hand and other DACCO equipment described in the literature showed promise (as did equipment from other potential system suppliers). The customer therefore asked the DACCO sales engineer to bring an applications engineer from the factory to look over the requirement for data collection and make recommendations.

After participating in preliminary discussions at their meeting, our applications engineer drew a functional diagram. This diagram formed the basis for detailed discussions and gave all the participants the opportunity to define data points, throughput rates, and output formats. All of this helped the IRRA engineers. It was also extremely valuable to DACCO's engineers, giving them an opportunity to understand the system's requirement and inject some ideas into the IRRA plans. A significant detail was that there were several cases in which DACCO standard products were capable of performing the necessary functions in the system. In some of these cases the IRRA engineers chose to modify their requirements slightly to accept the DACCO units. This gave them the assurance that they were specifying a producible system and that many standard products were available which were existing, proven reliable, maintainable products, rather than new designs.

After this productive meeting, the DACCO applications engineer and sales engineer planned the follow-up. Every few weeks, as they analyzed the problem and prospective solutions, they would send additional comments and suggestions to IRRA.

As one would expect, the IRRA technical specification reflected many of the DACCO ideas and products. This was appropriate, since IRRA recognized their value in the overall planning sequence.

BID SET FROM THE CUSTOMER AND BID DECISION

When the official Request for Proposals (RFP) IR-99636-E-89 was received at DACCO, the technical specification defined the system requirement as the DACCO sales engineer and applications engineer had anticipated. A few other items were added to the bid set. These included the customer's terms and conditions, a set of representations and certifications to be completed by bidders, a set of proposal instructions, and a few pages of quality control, test, documentation, and service requirements.

We had defined our marketing strategy already, based on preliminary understanding, so the first action at DACCO was to examine the bid set and verify that it was essentially as we had anticipated. A meeting was held then to make the official bid decision and to reconsider and possibly fine tune the marketing strategy. The decision was to prepare and submit a proposal.

STRATEGY

We decided to propose a system that would be almost totally compliant but that would offer slight deviations in some areas in order to offer our standard products and save the expense of redesign. We would show the customer that the benefit of offering proven products with excellent documentation and service would outweigh the minor disadvantage of nonconformance. We recognized that this approach would allow our price to be significantly lower than if we attempted total conformance. We would not use this as an argument in the technical proposal, however, because price speaks for itself.

PREPARATION AND PROPOSAL KICKOFF

The first 48 hours after the bid decision were the busiest of the entire bid sequence for our proposal manager. He had to translate prior knowledge and marketing strategy into a package that would instruct and motivate the proposal team, get copies of the appropriate sections of the RFP for each person on the proposal team, take care of the many planning details, and convene a kickoff meeting.

Document 9.1 shows his Win Strategy handout for the team members. Document 9.2 shows his tentative proposal outline, with a suggested page budget for each section. The main purpose of this budget was to let team members know in broad terms the relative depth of discussion on each subject. His Storyboards 9.1 to 9.3 flush out several of the major paragraph headings. More detailed storyboards for the subparagraphs would follow, after the system architecture was defined further.

The kickoff meeting was conducted by the proposal manager, after brief opening statements by the president or division manager and marketing manager. The proposal manager was deputized by top management to make the necessary decisions and provide the leadership to win this job. He had the full support of management. Any failure to comply with his directions would be considered detrimental to the good of the company and handled appropriately.

The proposal manager described the customer's Rocket Test Stand and then outlined our required response to the requirement for a Data Acquisition System for this stand. He outlined the marketing strategy and led the team members through an examination of the bid material. At each stage, he verified that team members understood their roles in winning the upcoming contract with IRRA.

Finally, the proposal schedule (Document 9.3) outlined all of the activity necessary to meet the RFP due date. Note that a calendar form was used; as it is easier than a simple tabulation of dates.

Before the RFP arrived, our proposal manager had compiled a tentative Section I and Section II, and attached them to this package also to guide the team members. He attached also a sample work breakdown structure for this job (Document 9.4), and the cover letter (Document 9.5).

Document 9.1   Win Strategy
How We Will Win (or Lose) This Job

We are in a good position on this bid. The customer likes DACCO, and feels that we can provide the best system for him. If we can give him the proper ammunition, he will go to bat for us in the competitive procurement process.

We will *lose* if we try to sell the *XY* package, but we will *win* if we offer the *QQ* series. We will *lose* if we offer *LMN*, but we will win if we configure the *OPS* as he wants them. The logic behind this doesn't matter any longer (these are customer hang-ups that we cannot overcome)

We will win with a decent, understandable proposal and a reasonable (not necessarily lowest, but justifiable) price. My suggested proposal outline is suitable and I would be concerned about any major deviations from it. The proposal will be evaluated by a sympathetic engineering team. Justification of their recommendation will be easier and more certain if the proposal is readable (and rereadable, well indexed, and cross-referenced).

Our gain when we get this job is high. We get the follow on as defined by them. Also, we establish with the IRRA our capability to handle the test stand format, something we have never done before.

Please conform to the Proposal Outline as shown. Also, please do not make me a liar in my Section 1 and Section 2 (attached). If we can back up these two sections with a reasonably good proposal at that previously mentioned justifiable price, the job is ours.

(Signed)

Proposal Manager

Document 9.2
Proposal Outline for Rocket Test Stand System

SECTION 1: INTRODUCTION
(*writer's name in parentheses*) 4 pages
Discuss the unique features which we offer.

If a page budget is appropriate, add it to the outline.
It lets the team members know the relative importance of each
section. The budget is not firm, but simply a point of departure.
Don't overdo the Introduction section.

SECTION 2: SYSTEM DESCRIPTION
(_____) 6 pages
Give an overview to show the generic data stream. Define
inputs and outputs. List the quantities of equipment and services. Show a simplified data flow diagram.

Define the general subject matter to be covered in each
section.

SECTION 3: EQUIPMENT DESCRIPTION
3.1 Unit 1
       3.1.1 Generic
       (_____) 4 pages
       3.1.2 Specific modifications
       (_____) 4 pages
3.2 Unit 2
       3.2.1 Generic
       (_____) 6 pages
       3.2.2 Specific modifications
       (_____) 2 pages
3.3 Computer Interfaces
Show format for data transfer, logic levels, handshake,
and any other characteristics for the output interfaces on
each stream.

For each subsection in the Equipment Descriptions, define
the general topics which must be addressed to influence and
satisfy the Customer.

3.6 System Control
(_____) 2 pages
Show the computer access to all setup lines.

3.7 Patching, Cabling, Rack Mounting
(_____) 3 pages
Tell about cables.
Talk about rack size and nontilt bases.

Don't forget the little details that show your total under-standing of the customer's problem.

3.8 Input Power, Environment, EMI and such details
(_____) 3 pages

Even power, environment, and such details deserve some comments.

SECTION 4: PROGRAM MANAGEMENT
(_____) 20 pages
Show how we will run the job, and include good up-to-date resumes on several of the key people. Talk about quality assurance also.

Your performance on this job is of primary concern. Don't let it get lost in the boilerplate.

SECTION 5: SPECIFICATION COMPLIANCE
(_____) 5 pages
Provide a paragraph by paragraph list of cross-references between their specification and our proposal. For any exceptions or clarifications, show how we chose them for the customer's benefit.

SECTION 6: CORPORATE CAPABILITIES
(_____) 20 pages
Show our general boiler plate on this subject, but flavor it with as much detail as practical.

Points That Will Impress IRRA the Most

1. We are experienced in handling _____ _____.

2. We are leaders in _____.

3. Our _____ is high speed, state of the art, with diagnostics.

4. Field service, training and other support services are available from us.

5. Our system is expandable by the addition of other streams.

This is where you develop your "win strategy"

Possible Weakness

1. Proposal of a lower-cost system by someone who may take exception to some other specifications. Our best defense against this is a cost-effective bid, with minimum deviation from specifications. They should be able to justify our selection based on that set of conditions.

Lay the facts on the line here. Every team member should know your weaknesses in order to combat them effectively.

Some Terminology To Be Used Throughout the Proposal

1. Our Name   *DACCO*
2. Customer's Name   *IRRA*
3. Overall System   *Rocket Test Stand*
4. Our Subsystem   *Data Acquisition System*

Storyboard 9.1

Proposal Paragraph: *3.1.1*
Subject: *Bit Synchronizer*
RFP Reference: *3.1.1.9*
Subject: *same*
Requirement: *Operation to 5 megabits per second, eight PCM codes. Ability to make bit decisions to within 1 dB of theoretical accuracy curve.*

Our Bid Strategy: *Bid DACCO 720 and emphasize conformance in proven unit.*

Theme sentence for this paragraph: *Proposed Bit Synchronizer meets or exceeds all specifications, and is a proven design from DACCO.*

**Subheadings
and emphasis**

Features

Description

Illustrations

Photo of front panel
Block diagram, simplified
Specification tables:
Inputs
Synchronization
Displays
Performance curves

Page budget: *10* (double spaced).
Figures: *4*
Tables: *3*

Comments: _____
_____
_____
_____
_____
_____

Preparer: *Brubeck*   Phone: *5678*

Storyboard 9.2

Proposal Paragraph: *3.1.2*
Subject: *Digital/Analog Converter*
RFP Reference: *3.1.1.8*
Subject: *same*
Requirement: *16 DACs, 64 discretes in base unit, expandable in field to at least 64 DACs and 128 discrete total. Resolution: 12 bits*
Our Bid Strategy: *Use standard 8350, which exceeds specifications.*

Theme sentence for this paragraph: *Proposed DAC/Discrete unit meets or exceeds IRRA specifications, and is a standard DACCO product.*

Subheadings and emphasis	**Illustrations**
Features	Front panel
	Block diagram, simplified
Description	Table of specs:
	Base unit
	Word selector
	DAC
	Discrete

Page budget: *10* (double spaced).
Figures: *2*
Tables: *4*

Comments: _____
_____
_____
_____
_____

Preparer: *Brown*   Phone: *6619*

Storyboard 9.3

Proposal Paragraph: *3.2*
Subject: *System Software*
RFP Reference: *3.2.4*
Subject: *same*
Requirement: *Setup, acquisition, display, and storage. Easy to use. Based on standard operating system and compiler.*

Our Bid Strategy: *Use DACCO standard software, and develop new display package to meet spec details.*

Theme sentence for this paragraph: *The proposed software meets specifications, and is modular and expandable.*

Subheadings and emphasis	**Illustrations**
Database	Current value table
Acquisition and storage	Dual buffers, Display
Print software	Print disk to DAC
	Software disk to DACs

Page budget: *18* (double spaced).
Figures: *1*
Tables: *0*

Comments: _____

_____

_____

_____

_____

Preparer: *Stone*   Phone: *6699*

Document 9.3   Proposal Schedule

JUN 19	20 Receive Bid set from IRRA	21 Bid Decision Team Definition	22 KICKOFF Meeting 1 PM	23	24	25
26	27	28 Firm definition of system architecture 1 PM	29	30	1	2
JUL 3	4 Holiday	5	6 Progress review meeting 1 PM	7 Sketches to Graphics Dept by noon	8 Tentative write-ups in by noon	9
10	11 Costs to accounting by noon	12	13 Proposal changes due by 1 PM	14 Print, bind proposal / Pricing decision	15 Mail via Fed Express 3 PM	16
17	18 Due at IRRA 4 PM					

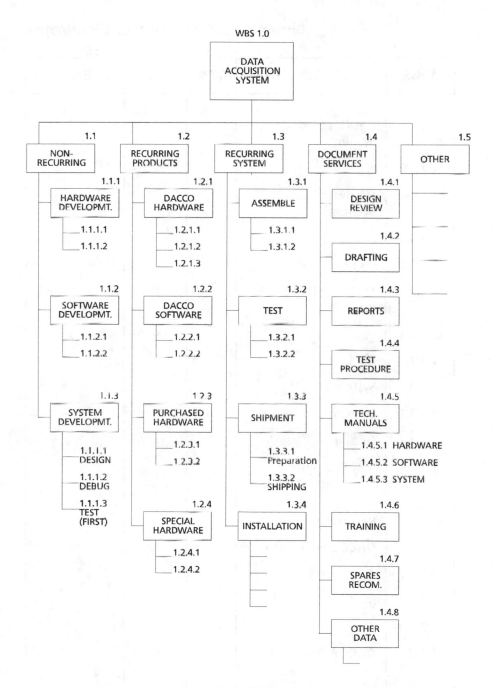

Figure 9.1   Sample work breakdown structure (WBS)

Document 9.4   Cost-Estimating Matrix

WBS	DESCRIPTION	QTY	EACH	EXTENSION
1.0	Data Acquisition System			
1.1	Non-recurring			
1.1.1	Hardware Development			
1.1.1.1	A			
1.1.1.2	B			
1.1.2	Software Development			
1.1.2.1	A			
1.1.2.2	B			
1.1.3	System Development			
1.1.3.1	Design			
1.1.3.2	Debug			
1.1.3.3	Test First System			
1.2	Recurring, Products			
1.2.1	DACCO Hardware			
1.2.1.1	A			
1.2.1.2	B			
1.2.1.3	C			
1.2.2	DACCO Software			

WBS	DESCRIPTION	QTY	EACH	EXTENSION
1.2.2.1	A			
1.2.2.2	B			
1.2.3	Purchased Hardware			
1.2.3.1	A			
1.2.3.2	B			
1.2.4	Special Hardware			
1.2.4.2	B			
1.3	Recurring, Per System			
1.3.1	Assembly			
1.3.1.1				
1.3.1.2				
1.3.2	Test			
1.3.2.1				
1.3.3	Shipment			
1.3.3.1	Preparation			
1.3.3.2	Shipping			
1.3.4	Installation			
1.3.4.1				
1.3.4.2				
1.3.4.3				

WBS	DESCRIPTION	QTY	EACH	EXTENSION
1.4	Documentation			
1.4.1	Design Review			
1.4.2	Drafting			
1.4.3	Reports for the Customer			
1.4.4	Test Procedure			
1.4.5	Technical Manuals			
1.4.5.1	Hardware Products			
1.4.5.2	Software			
1.4.5.3	System			
1.4.6	Training			
1.4.7	Spares Recommendations			
1.4.8	Other Data			
1.5	Other			
1.5.1				
1.5.2				
1.5.3				
1.5.4				
1.5.5				

Document 9.5   Cover Letter

(Date)

Gentlemen:

It is a pleasure to present this proposal to the Interplanetary Rocket Research Agency in response to your Request for Proposals number IR99636-E-89 dated _____(with amendments _____and ____).

We have analyzed your requirements for this equipment to be used on your Rocket Test Stand, and the turnkey system we propose is ideally suited to your present needs. In addition, we offer many features that ensure growth potential, so that the system can be expanded if you ever need such expansion.

Some of the strong points in our proposal are:

* _____
* _____
* _____
* _____

I hereby certify that _____
_____
_____

Use the "certification" comment if it is required by the Request for Proposals, or attach the required certifications if they are lengthy.

We are fully committed to providing this equipment if we are selected. This is my personal assurance that the Program Manager and his team will operate with the full resources of the company to satisfy your needs in a timely manner.

We will be pleased to have IRRA personnel visit DACCO during the evaluation of this proposal, and/or visit your offices for formal presentations or informal discussions. Please contact our Sales Engineer in your area, _____, or our Proposal Manager, _____, to set up such a visit or to resolve any points that require elaboration.

(cont'd)

I am sure that this proposal meets your requirements in a fully compliant manner. Our company looks forward to your award of the contract for this important equipment.

Sincerely,

Vice President or President

Attachments: Proposal Abstract
Technical Proposal (_____copies)
Price Proposal (_____copies)

TECHNICAL PROPOSAL

As assigned, DACCO engineers and programmers worked with the proposal manager to establish a system architecture. Each then wrote the appropriate sections of the proposal. The basis of each major paragraph or subparagraph was the relevant storyboard as handed out at the kickoff meeting and/or subsequent detailed discussion times. See Chapter 13 for major excerpts from the various sections in a sample proposal.

PRESENTATION

IRRA evaluated their proposals, and invited DACCO (among others) to make a formal presentation of 2 hours duration.

The presentation team DACCO selected included:

1. The sales engineer who covered the IRRA account (because he knew the people, and had established rapport).
2. The general manager (because he could speak about DACCO's commitment to the proposed contract).
3. The proposal manager (because he was a good speaker, and understood the proposal from beginning to end).
4. The systems engineer and systems programmer (because they could answer any questions on technical subjects).

The presentation consisted of:

1. Introduction by sales engineer (10 minutes)
2. Statement of commitment by general manager (10 minutes)
3. Proposal presentation by proposal manager (1 hour)
4. Questions and answers, chaired by proposal manager (40 minutes)

AWARD

We were notified by telephone of our selection, and the contract arrived 2 days later.

Our program team started work immediately, using the technical proposal and work breakdown structure as planning docu-

Document 9.6 Post-Award Analysis

Customer: *Interplanetary Rocket Research Agency (IRRA):*
Program: *Rocket Test Stand Program*

Competitors:
1. Most formidable   *Monster Systems, Inc.*
2. Next most formidable   *Orientronics, Inc.*
3. Next most formidable   _____
4. Next most formidable   _____
5. Next most formidable   _____

Brief description of the product(s) or system(s):
High-Speed data collection, processing, and display system

Quality of our rapport with customer 8   (0-10)

Quality of our past dealings with customer 9   (0-10)

Quality of our proposal 8   (0-10)

Prices (if known):
1. Ours   *$592,000*
2. *Monster*   *$665,000*
3. *Orientronics*   *$520,000*
4. _____ $ _____
5. _____ $ _____

Our significant positive features:
Standard hardware and software are almost totally applicable

Our significant negative factors:
No large system experience with this customer

Other observations:
*Customer is budget-conscious, but does not always go with
low bidder if other factors are significant.*

Result of customer evaluation:
Best technically: *DACCO*
Most attractive price: *Orientronics*

Winner: *DACCO*
Biggest reason: *Availability of standard products to do the job,
thereby holding price to a reasonable level.*

Improvements we should make as a result of this analysis:
*Expand our standard display software to include the special
items we develop for this IRRA project.*

ments. The sales engineer and program manager visited the customer the first week to say "Thank you" and to open the way for communications during the performance of the contract.

POST-AWARD ANALYSIS

After receiving notification of our selection as supplier of the Data Acquisition System, we compiled the post-award analysis (Document 9.6) for our use in business planning. We complete this analysis in any case (whether a win or lose) and distribute it to all persons involved in bidding.

10

Product Planning and Marketing

PRODUCT PLANNING

A CENTURY AGO, RALPH Waldo Emerson was credited with the statement: "If a man... make a better mousetrap...the world will make a beaten path to his door." While Emerson was a successful essayist, critic, poet, orator, and popular philosopher, the record shows that he never built a mousetrap, nor steam engine, telegraph system, automobile, nor television set. Records of the inventors of the past and present will show, too, that Mr. Emerson was wrong in his estimation of the public's buying habits. It is sad but true that the world does not "make a beaten path" to the door of an inventor. The American Management Association reports that 9 of 10 new products placed on the market do not win customer acceptance. The 1 in 10 that succeeds does so because of a well-planned design and marketing program. The "beaten path" is

119

made by the seller and not the buyer. This ratio was even applicable in the days of Thomas A. Edison, who patented over 1000 devices and found a market for about 100 of them.

Causes of New Product Failures

The Small Business Administration notes the following major causes of new product failures:

1. Failure to pretest products or packages with distributors and consumers.
2. Lack of a well-thought-out marketing program.
3. Failure to test the market.
4. Insufficient product research (placing the product on the market before the bugs are out of it).
5. Inexperienced management.
6. Inadequate financing.

QUESTIONS TO ANSWER ABOUT THE PRODUCT

The following sections outline the considerations that must be made in introducing a new product and examine the following questions to be asked when any new product is developed and introduced:

1. Is such a product needed?
2. Is this a practical product for our company to develop?
3. What specific details will make the product salable?
4. What must we do before we deliver the first unit to a customer?
5. How should we sell the product?
6. How should the product be serviced?

Is Such a Product Needed?

Ralph Waldo Emerson thought the product was a better mousetrap. The answer concerning your proposed product is more likely to be based on results of the following analysis:

1. Can we improve on the size, weight, or performance of an existing product?

2. Can we build a product with equivalent performance to those on the market which will sell for less?

3. Have recent developments in the marketplace (an increase in recreational travel, for example) created a need for the new product?

4. Does a recent technological development (a new type of computer, for example) open the door to new peripheral devices?

5. Or (least likely of all) is this a unique breakthrough, comparable to the steam engine, the telegraph, the vacuum tube, or the transistor?

If the answer to one of these questions is "yes", the product has passed the first hurdle. If, on the other hand, the product is based on a "me too" approach and the hope that a strong sales effort can outdo the competition, the company may be headed for trouble.

Color television development in the 1960s and 1970s, for example, fits into these first two categories. Several companies became successful, and others lost their shirts, in attempting to answer these needs. Ease of maintenance, a smaller set, larger picture tubes, rectangular tubes, higher reliability, and lower price proved to be the key factors for the television marketplace. Other factors were less important, and those products failed.

In this analysis, consider the broad questions: Who makes a similar product? What are the advantages and the disadvantages? What would be the advantages of your product, and the disadvantages? What type of person buys the product? Where does the money to buy the product come from?

After a detailed and objective analysis, stand back and take a look. The biggest pitfall in this or any other phase of new product introduction is "playing a hunch," or making a decision as one would bet on a horse. The Ford Motor Company is reported to have bet $250,000,000 on the Edsel—and lost. Can your company afford that kind of bet?

Is This a Practical Product for the Company to Develop?

The answer to this question may be obvious, if a construction company engineer designs a new vacuum cleaner. The logical course of action is to sell it to a vacuum cleaner company for development and sale. On the other hand, the answer may be difficult to reach in another company's case. Some questions to be answered are:

1. Would this product fit into your overall corporate objectives?
2. Can it be produced by the personnel and the equipment and facilities currently available?
3. Can it be sold through your present sales organization?
4. Can you obtain the necessary patents or licensing agreements?
5. Can you obtain the necessary supplies to support your production line?
6. Do you have or can you readily obtain the finances required to develop and produce the product?
7. Will the production and sale of this product be compatible with the objectives of your existing suppliers, customers, employees, investors, and management?

A "No" answer to any of these questions is a red flag of potential danger.

What Specific Details Will Make the Product Salable?

We come now to the nitty gritty of market research. Having determined that the world needs the product, and that the company is capable of developing and producing it, what can be done to make it exactly what the world wants?

Consider the environment in which the product will be used. Is it to be used only with certain kinds of equipment? Then don't expect the eventual customer to rewire, or repaint, or relocate his other equipment to match this product. That won't happen.

Is it visible when used? Then don't expect the customers to ignore the esthetics of the product. That won't happen.

Is the price of this kind of product fairly well established in the marketplace? Then don't expect the customer to pay more for the product unless it has clear and obvious advantages that are worth more to the customer. That won't happen.

The conclusions should be obvious. Design the product to complement the customer's application, to be displayed favorably and look good, and to be marketed in the manner that suits the customer's financial needs. Otherwise, the effort is a waste of time and money. A good market research organization can be instrumental in providing an objective look at these factors.

Let's take a longer look at the optimum price for a product. Ideally, one should construct a curve showing that if the product is sold for 1.0 x dollars, selling so many thousand per year will result in a total yearly profit of so much. If it is sold at 0.9 x dollars, the sales will likely be much higher but what will the total yearly profit be? If you sell at 1.1 x dollars, the sales will likely be much lower, and therefore, what will the total yearly profit be? With such a curve, one can find an annual volume that is practical for the production facility, at a price that is fair to the customer and to the company. This can only be an educated guess, but it's the best basis for planning.

There will probably be several versions of the product or several optional features. Try to anticipate these as soon as is practical. Define each in a manner that can be understood by the market researcher and the customer. Price each version or option.

The product name can be a big factor in its sales appeal. Here again, one can obtain the services of advertising experts in choosing a name. Every advertising person dreams of the opportunity to introduce another Bird's Eye to the consumer, or another Sanka, Jell-O, Bisquick, Life Savers, Elsie the Cow, Four Roses, Lucky Strike, 20 Mule Team, Cutex, Dixie Cups, Kodak, or Teletype (all registered trade names). Why not dress up a good product with a good name? One deserves the other!

What Must We Do before We Deliver the First Unit to a Customer?

"We have the product designed; now go and sell it." When these words are first spoken, the product is probably about 30% of the way between conception and release to production. The breadboard works in the laboratory, but will it do what the world needs? Will the second one work, or were you lucky with component tolerances this time? Can it be simplified? Can it be packaged for production? After being packaged, will it work when subjected to the "real world" environment with vibration, heat, cold, humidity, and so many other hostile conditions?

These questions must be answered before one can be assured that the product is ready to be manufactured. The budget should be examined. Do the designers need more money before the sales bring in a return? A realistic schedule must be derived for all phases of development. This schedule should extend through the first 6 months of production. Significant milestones must be established and monitored. The company can't afford to start the actual sales effort too soon, or too late.

Is the product patentable? If so, the claim should be filed as soon as is practical.

Is documentation included in the schedule? The customer will not buy a product unless the customer furnishes operating instructions. Maintenance instructions will also be needed for use by the customer or the service organization.

Has the planner allowed adequate time for debugging after the first units come off the production line? Face it: problems will certainly present themselves in these first few units.

In summation, be realistic in scheduling activities which precede the delivery of your first units to the customer. Don't ask sales people to sell, and customers to buy, a "paper tiger."

How Should We Sell the Product?

Regardless of the size of the company, the type of product, the price, or any other general conditions, the probability of making sales without someone contacting the customer on his or her home turf is virtually nil (Ralph Waldo Emerson notwithstanding). A product must have a sales force.

A small company without a sales force has the option of enlisting sales representatives or selling through distributors. Sales representatives serve several companies, selling products for a commission (usually a percentage of the selling price). They have no responsibility for stocking units, nor for installing or maintaining the equipment after sale. Distributors, on the other hand, buy from you at a significant discount and sell to customers. They will usually install the equipment and keep a supply of spare parts for maintenance. The choice between representative and distributor is dependent on the type of product and the type of customer. The best bet is to examine each possibility in each area you hope to cover. Commissions and discounts are subject to negotiation. Keep in mind, of course, that this is the only sure way to motivate salespeople, so don't try to shortchange them. A company-established commission rate can be fair to everyone or unfair to everyone. If it is unfair, you will find out in a hurry!

Whatever sales organization you choose, it will be necessary to help the sales team in every practical way. They need training regarding your product but they don't need training in basic sales techniques! They need periodic written suggestions regarding potential customers for the product and new answers to customers' resistance. They need demonstration units to take into the customers' offices and operate in the customers' presence.

A well planned and properly scheduled advertising campaign is another essential part of new product introduction. Chapter 11 describes the basics of such a campaign.

One of the best ways to exhibit a product is at trade shows. A good trade show attracts several hundred to several thousand attendees. To see this many people in their offices would obviously be better, but it would take several salespeople all year long to do it. Don't miss taking a new product to the appropriate shows. Here again, get the benefit of an expert in planning the display.

Don't overlook the possibility that this product lends itself to the presentation of a technical paper. This may be one of the more difficult tasks for an engineer, but the impact is worth the effort. A technical article in a trade magazine is worth more than the same amount of advertising space. Chapter 12 discusses this form of publicity.

Is the company willing to loan units to potential customers for evaluation? This is a wonderful way to get more mileage out of the first production units. The sales department can send units on a memorandum billing for 30 days, to be returned or purchased at the end of that period. If the customer likes it, you have a sale!

How Should the Product Be Serviced?

Fallacies abound concerning the necessity for a well-defined field-service plan for a new product. The first fallacy is that one can convince the customers that the reliability of the product is so good that service is unnecessary. The truth is that the reliability is not that good and, even if it is, the customers will never believe it. The second fallacy is that if the customers have any problems, they can return the units to the factory for repair. The truth is that the customer has no assurance that your company will repair and return the unit promptly (after all, they have been dealing with other repair workers for years). The third fallacy is that customers don't think about such things before they buy, but the truth is that this is one of the first considerations of the typical customer.

The large company with an established field service organization must bring that organization up to speed on a new product. This involves having available operation and maintenance instructions, spare parts in each service office, and the assurance of instant support from the factory on major problems. For the company without a service organization, the same benefits can sometimes be offered by one of the several professional nationwide service organizations. These companies need training, spares, and factory support and, of course, they must be compensated for their services. In a typical arrangement, the manufacturer will pay them for installation and in-warranty repairs. They will contract with the user for out-of-warranty maintenance. The important thing is to make all of the necessary service arrangements before selling the product.

How Should Sales Literature Be Used?

How can you keep talking after you say "Good-bye"? The Sales Engineer just completed a productive hour-long visit with potential customers who listened carefully. They asked questions. They were quite obviously impressed, but how much of what was said will they remember on the day they write procurement specifications to issue a purchase order? Most of what was said will be forgotten in a few days.

Sales literature keeps talking after the salesman leaves the customer. Good literature emphasizes what was said, poor literature detracts from what was said, and no literature allows the customer to forget what was said. Good sales literature is the partner of the sales pitch. Don't leave home without it!

What constitutes a good piece of sales literature? A thorough description of the product should be written to take advantage of all the product's good features, admit its limitations, and appeal to the customer's reason and emotions.

To prepare good sales literature, look at the work of experts. Find brochures from companies in the $1 billion plus category who have full-time, highly paid professionals writing and producing their sales material. Incorporate their patterns and use a check list to ensure that you don't forget anything. The following characteristics make for useful and effective sales literature:

1. Product model number and a descriptive name.

2. A short, meaningful description that tells your reader whether or not to keep reading.

3. Several features, briefly stated but using quantitative terms. For example, not just "high accuracy," but "less than 0.1% error."

4. Product photograph (so the customer knows it really exists).

5. Input characteristics—all of them.

6. Output characteristics—all of them.

7. An explanation of what happens between the input and output, and some applications or examples.

8. Controls, indicators, and displays (this could include a blown up photo if you can get some mileage out of it). Also, does the unit have remote controls, or indicators?

9. Operating electrical power such as voltage and acceptable range, line frequency and range, and wattage under typical conditions.

10. Environmental characteristics, especially temperature range.

11. Dimensions, weight, and mounting provisions.

12. Ordering information. A list of components in the basic package (accessories, cables, technical manual).

13. Options, accessories, and supplies of general interest. How to order each of them.

14. The company name and logo, address, phone number, Fax and Telex numbers. Possibly a space for the salesman to stamp his or her name and phone number.

15. Reference to related equipment, and to upgrades (but not less expensive devices). Here is an opportunity to expand the market with this customer.

16. Date of issue. The customer needs to know when your literature should be updated.

Examine a sales brochure carefully before it goes to press: does it tell the reader why he or she should buy from you? If not, stop the presses until you revise! Be alert to the supply of sales literature, and reorder when the bin is low. (Brochures reproduced on the copier look terrible!).

CONDENSED CATALOG

Your first step in high-technology sales is customer education. If a company or division has more than a dozen different but related products, a Condensed Catalog is needed. This gives the sales force something from which to talk. It gives the customer who is "not interested now" a reference for the files. It provides a mailer for general inquiries. It may even provide a paid insert for a Buyer's Guide or other directory.

In preparing material for a catalog, one must leave out lots of good information. This is difficult! Each product should occupy only enough space for prospective customers to know whether or not they are interested. If the company has similar but not identical products to serve different applications, do the reader a favor by using brief comparison charts in the catalog. If the company has a product, or group of products, whose application is not generally understood, provide a half-page or one-page primer for the reader.

Be careful how many catalogs you print. (Don't fall for the printer's pitch that the second 5,000 will cost almost nothing.) The company will not revise the catalog until the entire printing is used. This means you can't introduce the newest and most exciting products until it's time to print a new catalog.

APPLICATIONS NOTES
AND OTHER TECHNICAL AIDS

Isn't it frustrating to be limited in the size of a sales brochure? Wouldn't it be easier just to hand out the technical manual on a new product to every potential user? Think of all the questions it would answer! The problem is, the very size of a technical manual is threatening. Someone who may not know if he or she needs the product will never pick up the manual and read far enough to see what the product can do. You can supply technical details to the potential user in stages:

1. One page advertisement, mailer, sheet in the condensed catalog, or other introductory material.

2. More detailed technical information in the sales brochure for the person whose "possible interest" is stimulated.

3. Detailed applications information, graphs, interface descriptions, for the person who sees "probable interest" after reading the sales brochure. A special *Applications Note* on each complex product answers those final questions before a purchase decision is made.

4. The technical manual when the equipment is delivered, or perhaps in advance if it is needed to close the sale.

An Applications Note tells a potential user almost everything about a product that is not proprietary and that relates to use of the product in the typical range of applications. It is the engineer-to-engineer communication device, edited by the marketing professional but essentially the "whole story" for a potential user. It tells exactly what the interfaces are to the outside world—not just generally but down to the timing of handshakes. It tells exactly what the product does under typical operating conditions—not just an overview but with charts and graphs and tables. It is the insurance that, if customers had any remaining questions after reading the sales brochures, they now have the answers.

The Applications Note doesn't always "close" a sale. Sometimes it tells a potential customer that this product is inappropriate for the application. Isn't it best to learn this before making the purchase? Better to lose a sale on one product than lose credibility forever with a good customer!

Don't forget, of course, that a surprise in the Applications Note is not necessarily a lost sale. It can be the opportunity to offer a minor modification at a reasonable increase in price to overcome the difficulty. It can be the opportunity to talk the customer up to another product in your catalog that sells for twice as much and can do a lot more. The Applications Note is a useful tool when written properly and used properly.

Make Your Price List a Useful Tool

Can one of your company's sales people stand in a phone booth and take a simple order for a standard product? Can a sales secretary in a field office give fairly responsible answers to a customer's inquiries? Can one of the applications engineers talk to a customer or salesperson by referring to only one book? The answer to all of these questions is usually "no" but it could be "yes."

A price list should be a useful tool, not just a dull set of numbers. It requires a few hours of work for one person to produce, but saves many hours per year for several persons. Price quotations, technical details, and various categories of availability can be obtained from one source.

To create a useful price list, start with model numbers (arranged numerically) with prices. Add the ordering code derivation (shown graphically) and show ranges for all items on the code (minimum and maximum frequency, types of filter, and so on). Add comments liberally. Cover all the questions you had the first time you saw this item on a price list. How many modules can the rack adapter hold? How much rack space does it require? What else is required to operate this unit? Show it clearly. What is included in the base unit and price? Many customers forget to order spares kits, and would appreciate a reminder of the availability and price. Tack that onto the price list after the listing for each product.

Finally, options can represent a significant amount of business. What options are available? Why is this one or that one particularly useful? What is the price of each?

The customer is likely to be working from a condensed catalog, so incorporate the excerpt from this catalog that relates to each product. For the final go-through, add other comments as appropriate. Remember, this is the "bible" for real time dealings with your customers. Make it a useful tool for expanding your sales!

Apply each of these questions to each product in your price list:

1. Is the complete ordering code shown, so that a user can derive an order from this one sheet? Are prices shown?

2. Is the physical configuration defined in general terms? If the product is rack mounted, how high is it? If it is modular, what does it plug into? How does a user order that device? How many modules can be plugged into a single device? What else is required for the product's basic operation?

3. Are all the commonly available options shown, with prices? Is the value of each spelled out clearly?

4. Is a spares kit shown on the list?

5. Does the unit have expendables? Are they shown and priced?

CONCLUSION

Let us reword Emerson's 1871 statement to reflect the marketplace more realistically.

"Even though a man makes a better mousetrap, he must still beat a path to the door of the customer."

Example 10-1. Wacky had a design group leader who was extremely creative (good), an adept circuit designer (good), and a strong team leader (good), but he was unwilling to listen to marketing (uh-oh!). This engineer was designing a new product with given operating characteristics to meet a very clear need in the technical field in which the company was a perceived leader. This product would have an output current of 0–25 milliamperes, and in some applications it would drive a pen recorder through some preamplifiers that were standard on the recorder. One day, in a moment of "inspiration," this engineer decided (unilaterally) to increase the output current to 0-500 milliamperes to eliminate the need for preamplifiers. The problem, however, was that output amplifiers made the unit (a) larger than the market could stand, (b) heavier than practical, (c) hotter than the blowers could handle, (d) too expensive for the market, and (e) incompatible with the pen recorder industry. It didn't sell.

Remember: Design with an open mind.

DEVELOPING THIS SUBJECT

1. You are the marketing manager of a minicomputer company. You do not now manufacture terminals for your computers, because you have several good sources for these devices. What developments can you envision that would make it wise for you to start manufacturing your own terminals?

2. You have a high-technology device in your home or office that you know well and like. Prepare the basic details that should be used in a sales specification sheet for this device. If you are in a study group, critique the specification sheets of the other participants.

3. Develop a quarter-page synopsis of this device for use in your company's condensed catalog. Critique the other synopses from the members of your study group.

11

Advertising and Sales Promotion

D O YOU SELL YOUR SOUL
to the devil when you advertise?

MAGAZINE OR "SPACE" ADVERTISING— A BLESSING OR A CURSE?

To many engineers even the word "advertising" is anathema, that is, something shady and indulged in only by charlatans. The reason for this attitude is obvious as every day we are bombarded by consumer advertising everywhere we look. But there is consumer advertising and then there is industrial advertising! The former is largely showmanship. The latter is completely different.

The advertising in an industrial magazine ("space" advertising) is aimed at engineers and scientists and their associates. These people have been trained to be accurate, practical, conscientious, and honest. Why? Because their work demands it. You can't lie about a new phase-locked-loop circuit if it doesn't meet specifications. Nothing you can say will change that fact. The larger the system, the easier it is to hide defects from untutored observers, but in the end the defects eventually will surface. Engineers and other technical people aren't dumb! They've spent their best years pursuing academic goals that are difficult to obtain, requiring hours of study, dedication and, above all, intelligence. When a company advertises to them, it must keep that in mind.

Industrial Product Advertising—How to Do It

If you think that all advertising is the same, flip through a technical magazine and look carefully at the ads. Ask yourself: Would that ad persuade me to find out more about that product? Would I even pause to read it? Do I know what product they're advertising and what it does? If you can answer any or all of those questions positively about any of the ads, you are looking at an effective industrial ad. With industrial advertising in trade magazines, there is an old adage that applies to the kind of ads that should be run: "Oh tell me quick and tell me true, or else my dear, the hell with you!"

The essence of industrial advertising is to say right up front whatever you have to say about the product. Otherwise, busy engineers, flipping through 200 or 300 pages of their favorite magazines, aren't even going to pause. A headline that says, "You can rely on Acme Chemicals for the best products," or "We'd like to blow our horn about our new product," or "WACKY does it again" isn't going to slow them down. If the engineers were asked for their first reaction to such an ad, it would probably be "Who cares? I haven't got time to read through every silly ad to find out what they're selling and what it does!"

The type of headline that grabs their attention will tell them immediately what you've got to sell and what a great product it is, "40 Gigahertz in a transmitter you can carry in your pocket," or "An all-purpose missile that can turn on a dime at Mach 2," or "Microchips you can program yourself!"

Examine the features of the products you're selling and select the one or two (no more!) of each that beat out the competition. Prepare a short headline that tells the reader what those features are. Now do it again with a different headline. And again. In fact, do it in as many different ways as you can think of. When you've exhausted your creativity, put the titles away for a day or two.

After that cooling off period, pull the headlines out and look at them again as if you were a stranger to the product. You'll be amazed at your reaction to some of the headlines you thought positively brilliant. Don't give up! Eventually, you will get some that seem to be OK. Keep the ones with possibilities. Now compile all the features of the product you want to convey to your potential customers. Remember that not all of them can be put into a single ad. Be selective.

Once you have pulled together this information, call your ad agency, or start the process of selecting an ad agency if you don't already have one. The selection of layout, type style, and illustrations for an effective ad requires an expertise seldom found outside an experienced agency. All the agency needs to create effective industrial ads is guidance as to what to say in the headline and the kind of information that should be in the body text.

If you have an agency, give them the headlines you've created and the features you've collected concerning the product. Don't be shy about telling them the points stressed here: they must be specific about the product and its features, you don't want "cute" headlines that don't tell the reader anything, and you want to grab the attention *quickly* of readers who are potential customers.

Then forget about "pride of authorship." The agency might very well create a headline of its own based on the ones given them. Most agencies hire intelligent people who are capable of creating effective industrial ads once they've gotten good direction. When they create headlines, they think in terms of the overall idea of the ad and how to present it in a pleasing and effective way. That may require a different approach than you took, so they may change the headline. Just be sure they don't lose the message in the process.

How Big Should an Advertising Budget Be?

A small advertising budget with great ads is just as ineffective as a big budget with bad ads. The literature is filled with opinions about what constitutes a proper industrial advertising budget for magazines. Some say it should be a percentage of the yearly gross. If the percentage remained constant, however, you'd be spending more for advertising when business was good than when business was bad, which is just the opposite of what common sense dictates. Others say that it depends on the market and the amount you expect to receive if the product captures a reasonable proportion of that market. Still others say that each individual product doesn't make all that much difference, as long as you have continuity in your advertising program so that people know who you are.

There's validity in each of these opinions, but nobody can say what a particular advertising budget should be because needs and markets are different. Some useful rules of thumb can be applied to help estimate the budget, but experience and careful analysis are required to make the final determination.

How big is the total market? Do you know almost every company that could possibly be interested in your product? If you do, it might make more sense for you to launch a direct mail campaign rather than a more expensive space advertising program. Don't make assumptions about this, however, without good marketing information to support them. No matter how stable markets appear to be, they are all subject to sudden shifts. Just ask the railroads, the buggy whip industry, and the slide rule manufacturers. You may know all the potential customers today, but don't count on that knowledge tomorrow.

If your company doesn't have a lock on the market, space advertising is one of the best ways to tap unknown customers. Select the most widely read magazines in the field. Find out how much an ad in each will cost by calling the publishers directly (their phone numbers are on the mast head) or look them up in the Industrial Advertising volume of *Standard Rates and Data*, a monthly trade periodical found in most public libraries.

Study this "bible" of the industrial advertiser, because it is loaded with important information. For example, the magazines listed in it are organized according to markets. By looking up your market(s), you may find magazines you hadn't even heard about. For another thing, it lists the total readership of each magazine and the general categories of the readers' fields of interest.

For budgetary purposes, pick the magazines which are important and notice how much their ads cost. Ads of all sizes are listed; compare the prices for ads of various sizes to see how much it would cost to run an ad every month or every other month for a year. Remember, the larger the ad, the better the likelihood that it will be seen, but if you have to choose between, say, a half page ad every other month or a full page ad twice a year, pick frequency over size.

Once the company has decided on an ad schedule, decide how many different ads to run. To get the maximum benefit from an ad, run it at least three times to make it cost effective. Studies show that a good full page ad will be seen and noted by about 20% of the readers and an excellent one by 30% or more. Approximately the same percentage who missed it the first time will see it the second, and so on for each subsequent time that it runs. So for a 20% noted ad, almost half (48.8%) of the magazine's total readership will have noted the ad after it has run three times. For a 30% noted ad, about 65% will have noted it after three times.

Ask the agency how much it will cost to produce an ad of the type that you want, and multiply the cost by the total number of ads to get an approximate production cost. If you don't have an agency, ask several agencies for an approximate price, or call someone you know in the same business and ask them what they pay for ad production. Take the total amount that you have come up with, add a 10% fudge factor, and you've got an approximate space advertising budget to take to management for approval. Once a budget is submitted and approved, you're in a position to talk seriously to some ad agencies.

HOW SHOULD YOU PICK AN AGENCY?

A good rule of thumb for industrial advertisers is that the smaller the agency, the more likely it will provide good service at a reasonable price. Big agencies are expensive and they won't put their top professionals on an account that bills less than a couple of million or more a year. In fact, with an ad budget of less than $500,000, a big agency may not even take the account. If they do, they will assign their most junior people to it—people who are right out of school.

Try to find a number of good agencies in the region and invite them to make a presentation. You don't have to go to New York, Chicago, or Atlanta to get good advertising. In those advertising centers, most of the good advertising people are already working for the super agencies. You'll probably be surprised at the number of good, small agencies in your own region or even your own city or town. Most are composed of people who don't want the rat race of big-city advertising but whose talents and experience are on a par with some of the best. In any case, you are less likely to get anybody who's just graduated from school. A small agency seldom survives very long with junior talent.

The best way to determine if an agency is a good one is to see what they have done with a similar campaign, and then to see what they can do with one of your ads. Unfortunately, many agencies won't create an ad on speculation. Ads take time and a great deal of thought to create, and most small or medium-sized agencies can't afford to waste their time and talent on a presentation that may earn them nothing.

However, one can select four or five agencies who have shown promise in their initial presentations and tell each that you will pay them for the production of one ad to be used in a competition for the account. Although you will be spending money on two or three ads that will probably never see the light of day, it is money well spent to see what an agency can really do and how it will handle your account in the future. Give each of them the headlines you've created and the body text information, just as if you had already hired them.

It would also be informative to let the agencies charge their regular amount for the production of the ad. Tell them that the rate they charge will be what you expect from them for future ads if they get the account. That way you'll get a good idea of how expensive the agency will be. It will inform them that you are not planning to become their private Fort Knox if they become the agency of record.

WHAT SHOULD YOU EXPECT FROM ADVERTISING?

Done correctly, industrial space advertising can be a very powerful tool. You can use it to obtain new leads to customers, to persuade them to send for brochures and sales specs, and to inform potential customers in new markets how to solve their problems with your products.

You can never, ever, sell your product with an ad, alone.

Many companies make their mistake here when they advertise for the first time. They think that if the ad is a good one, it will sell their products, just like the soap companies sell more soap when they advertise. But these industrial products cost a great deal of money. That means that the customer has to know a great deal more about the product than you can put in a one page ad. Tiny 6-point type, printed margin to margin, just won't hack it. In many cases, the customer will need a salesperson to demonstrate the product before he can afford to take the financial risk of spending money on it.

An ad is the introduction to a sale—not the close.

If you want the inquiries to come in, offer free literature that will intrigue the reader. Before the start of a magazine ad campaign, have professional-looking literature ready to use to answer inquiries. A few typewritten sheets won't do.

The salesperson's rule of "Asking for the order" doesn't apply to industrial product advertising. Instead, ask the reader to send for literature on the product. Once he or she fills out that inquiry card, you've got a name that can be checked out and a company to call on.

The general purpose of space advertising for industrial firms is to generate leads and to keep the company name in the mind of your potential customers.

CHOOSING AN ADVERTISING
MEDIUM FOR YOUR COMPANY

Not advertising is like winking in the dark—*you* know what you're doing, but no one else does. When a product line is useful to professional customers involved in several different segments of technology, choice of the best advertising medium is not easy. A computer, for example, can be used by flight engineers, accountants, printed-circuit designers, and hundreds of other professionals. How can the manufacturer select specific target areas to sell the computer so as to address the market segments one at a time?

Standard Rates and Data gives every trade magazine's address, some details on contents, and the audited circulation figures (often, by category of reader).

Also, don't overlook the potential in buyers' guides published annually. The guides are year-round sales tools for you at no charge. All have free listings. Be sure that all your corporate and product names and trade names (past and present) are listed.

Example 11-1. WACKY has three divisions under the same roof. There are three product lines and three sales organizations but only one switchboard and one mail room. Each marketing manager decided to list his division's product line and sales offices or representatives in a national directory. Each asked for the forms, completed them, and returned them to the directory service. Good, so far! None of the three, however, notified the other two. This resulted in three entries, one after the other, in the directory, with conflicting information. How many employees—600, 700, or 1140? How many engineers—100, 200, or 290? Which division sells product A? product B? product C?

Remember: Coordinate directory entries, advertisements, condensed catalogs, and other general use information with counterparts in other divisions.

"FREE" ADVERTISING THROUGH NEW PRODUCT RELEASES

There is no such thing as a free lunch—but if you pay for one, it can certainly pull in a lot of customers. Many of the trade magazines have sections in which new products are announced to their readers. Any product deemed to have practical applications for the general readership and which is suitably described by a press release may be featured in the New Products section of a magazine.

Why does a magazine give this free publicity? There are two reasons. First, it is a service to the readers. Second, it may give the manufacturer of the new product a respect for the magazine and its readers (especially if several readers request more information and new market areas open up as a result), and make that manufacturer a good potential customer for paid advertising. Considering those reasons, then, how can one take advantage of this "free" service?

First, you must make a good choice of magazines. Second, the press release must be written in an acceptable format (length, general layout) for the magazine. And that's why it can't really be called "free" without the quotation marks. Certainly, it doesn't cost anything except printing and mailing costs to send the product release to the magazine, nor does the magazine charge if it decides to print the release. But that's where the rub comes in!

A good industrial magazine may get hundreds of releases every day. From those, a mere handful are selected for publication in any month. Which ones do you think will be chosen? There is a challenge for the responsible market specialist to describe in a well-chosen 80 or 100 words a fantastic product, which required 2 years of design time and sells for many dollars. Is it possible? Yes—but not easy! Is it worthwhile? Definitely!

Example 11-2. Company QQQ had developed a computerized monitor to measure the efficiency of large air conditioning systems. The company hired a one-person public relations (PR) company to write, produce, and mail a new product release. Company QQQ knew of only one market that could be interested in the product. That was original equipment manufacturers (OEMs) for air conditioners, because the product required system installation and was expensive.

The PR person was not hindered by that knowledge. There were at least six more markets that might be interested in the product—maintenance managers for large manufacturing plants, maintenance engineers for large hotels, restauranteurs with large walk-in refrigeration rooms, fast-food franchises, poultry processors, and meat-packing houses. For each of these markets, this specialist used *Standard Rates and Data* to choose the magazines reaching the largest proportion of the potential customers. A separate news release was created for each magazine, slanting the copy and features of interest to potential purchasers in each of the industries.

The result exceeded the expectations of even the PR professional. The releases were accepted for publication by a remarkable percentage of the 60 magazines that received copies. That part was not unexpected, but it didn't stop there! Five of the magazines thought the product release was so interesting that they ran a full article on it; two decided to do it in color, and one ran it as the cover feature article for the following month! The cost for the professional—$1,000 (including the cost of printing and mailing the piece). How much would you pay for the equivalent space advertising—over $100,000!

You can't buy advertising that's more effective than this type of coverage at any price! People will read an article much more readily than even a two-page color spread. In fact, a good article can capture 60% of a magazine's total readership. Compare that with the 20% or 30% for a good ad.

How can you pass up a bargain like that? Easy. When the PR specialist mentioned the fee, QQQ's marketing manager was shocked and said it was a waste of money! This was supposed to be free advertising, wasn't it? Fortunately, the owner of the company didn't agree.

IF YOU INSIST ON DOING IT YOURSELF

Nothing is more tempting to a marketer than to think he can get some free advertising without spending a dime. The temptation to do it yourself is almost impossible to resist. After all, the company makes lots of products. Why not try out your skill on one of them?

If you decide to give it a try, here are a few things you should consider. Remember what we said about space advertising. The people who read these magazines are busy (like the rest of us). They see several magazines each month and they know some material in each magazine is worthy of their reading time. When they thumb through the publication to see which 10%

they have time to read, they come to your press release. Will they read it? It's up to you. Can you attract their attention?

What would you say to a total stranger who doesn't know about your product, if you have some reason to believe he or she needs it? You meet the person in an elevator and have until you reach the 9th floor to convince him or her to stop and talk more with you and/or ask for a detailed specification sheet. Those first 80 to 100 words should be your press release, that is, the invitation to stop and listen to find out if he or she really needs the product.

Make the press release "you oriented." One needn't use the word "you," of course, but the whole message must point to the uniqueness of the product in a way the readers can grasp. There is no magic formula. Just keep all these things in mind as you write your well-chosen 80 to 100 words.

Print the release on the company's letterhead or on paper with the company's logo and address. In the upper right hand corner, type "New Product Release. For further Information, contact." and follow it with the name and address of the contact person. At the top of the release, type, "FOR IMMEDIATE RELEASE." Include a centered title similar to those on releases printed in the magazines for which you are writing the release. Start the first paragraph with a dateline. Finish with "END" so that the editor will know that there are no more pages to the release.

And for heaven's sake, include a 4" by 5" black and white photograph of the product or system with a caption attached to the bottom, including the product name and company name, so it can be identified if it becomes separated from the release. Don't send color photos unless the editor asks for them. Black and white photos can be easily copied by the publisher to include with the printed release, but color prints don't copy easily or economically.

Mail the release in an envelope big enough to hold the photo without bending. Address the envelopes to the "New Product Release Editor" of each magazine which may be interested.

Finally, follow the example of the PR specialist in Example 11-2. From *Standard Rates and Data*, select markets that could possibly be interested in this product. Be creative. Don't only think of the old markets. It won't cost much more to tap borderline markets with press releases.

FOLLOW-UP

So you have advertised your product, or sent new product releases to some magazines, or participated in a trade show. You have received or will soon receive request-for-information bingo cards, phone calls, letters, or other responses requesting more information. How do you prepare to answer these responses? You will torpedo the advertising program if you fail to follow up properly.

There are two extremes to avoid. The first can be insulting to the customer: sending something they have already seen or, equally useless, sending redundant literature. The second, sending the names of each respondent to call or visit, is a foolish waste of your salesperson's time. Each of us has seen the results of these responses. The first one cancels the whole effect of a well-planned advertising program. The second shows at least 80% of the respondents to be college students, absent-minded engineers who check everything and forget immediately what they asked for, or competitors (using home addresses, usually).

Avoid these extremes by taking the middle road. As soon as you have generated any advertising action, prepare a courteous and enthusiastic "form letter." Find or prepare technical brochures or applications notes which have new information in them to be mailed with the letter. Finally, prepare a "no postage required" reply card on which the recipient is asked to indicate:

1. Desire for some other piece of technical data which you are ready to provide.
2. Desire for your condensed catalog.
3. Desire for details on some related item of hardware or software which you make.
4. Request for an applications engineer or sales engineer to phone or visit.
5. Need for a demonstration at his or her office.

6. Desire to visit your plant for more detailed discussions.

7. Other (space for comments).

Probably 80% of the persons who receive this word processor-original letter, brochure, and card will file them or throw them away. Those people didn't really have a need. The others responding via the card, however, will be "twice qualified" potential customers.

You must follow up on the twice-qualified sales leads in an appropriate manner. If a request is made for literature, send it. If the potential customer wants a phone call or visit, make it. If it's a visit to your plant, set it up! Prepare a secretary or administrator to keep records and get a report when each twice-qualified lead is answered. Set up a system to notify the sales manager twice a month of all those leads that have aged for 2 weeks without appropriate action.

Do all of this before the first bingo card comes in!

Example 11-3. Ready-Fire-Aim. Our friends at WACKY decided to publicize their new Gizmo. They prepared new product releases and even bought some advertising space. Potential customers liked what they saw and wanted to hear more. They responded with bingo cards by the score.

What happened when they received these responses? They didn't have sales literature ready. Each respondent received a nice color copy of the advertisement to which they had responded! They were disappointed, to say the least, and no sales were made for a very good product. Remember: aim before firing!

Example 11-4. After the publication of new product releases and paid advertisements in various magazines, the president decides to have his salespeople follow up in person on the bingo responses.

The result: a few college students, a couple of engineers who didn't remember the articles, four or five bona fide prospects, and about $20,000 cost for WACKY on airline tickets and long-distance calls.

Remember: Qualify your responses before you send your salespersons knocking on doors.

TRADE SHOWS

For a typical high-technology business, one or more annual trade shows give you the opportunity to present your products and meet prospective customers. The technical conference should be the most meaningful week of your year. Don't waste it. How do you get the most mileage from the thousands of dollars you have decided to spend on exhibiting at a show?

When should you start planning for a show? The best time to start the plans for next year's show is the day before this year's show starts. That is when you begin to notice what mistakes you made this year. Keep a list, and you have begun to plan for next year. Did you forget power cords? or tools? or literature? or something even more important? Don't trust your memory—start a list.

When the show starts, watch the reaction of attendees as they approach your booth. Do most of them pause? Why? Of those who pause, what is their next move? Do they find the booth attractive? Inviting? Take a walk around the exhibit area. Stop and watch as attendees approach your competitors' booths. Who is attracting the most people? Why?

Don't be misled by the fact that your old-time customers all stop by and say hello. Is the main purpose of the show to say hello to the people you saw just a week ago? It is one of the purposes, of course, but you waste a lot of money if you fail to develop many new friends for the company during the show. What's the best way to accomplish this?

Your booth should be attractive, but not gaudy. Do your name and logo show prominently? Do you have two or three well-chosen emphasis points? Have you labeled your emphasis points, so that a timid stranger (the person with the million dollar contract) feels welcome to stop and look at the pride of your company? Neatly lettered signs let that person know what you are showing, so he or she can decide whether to stop and look.

Some of the displays should be dynamic. A dynamic display is captivating; the average person can't turn away until he or she sees what comes next—and next—and next. A graphic terminal, flashing menu, or sequence of projected photos grab the average person much better than glue. The booth where only the salesperson is dynamic doesn't fill the bill. People often feel intimidated, not attracted, by another human being whom they don't know.

Offer something to the casual observer by leaving a tray and a friendly TAKE ONE sign. It can be a one-sheet blurb on your newest system, a map of the town where the show is taking place, or a chart that is handy in your business, but offer something. By the act of taking it, the timid attendee (the person with the million bucks) has come a little closer to asking a question or accepting the invitation to stop and talk.

Make certain that your condensed catalog and your detailed sales literature are *not* easily available. This helps stimulate interpersonal contact, enables you to "qualify" customers, and tends to discourage competitors. Do have a large supply under a table in the booth area, however, easy to reach when the conversation reaches that point. Do yourself a favor by filing your literature in a file cabinet sized box with a rack and tabbed hanging folders. Then you can find a given specification sheet in 10 seconds, not 10 minutes.

The list for next year's show may include such items as:

1. Establish the roster of your attendees at least 60 days before the show, and make airline and hotel reservations.

2. Make all the appropriate arrangements with the exhibit manager at least 2 months in advance. Include such things as a carpet, chairs, or ash trays.

3. Get the display ready. Set it up in the plant for checkout. Ship it soon enough to guarantee arrival the week before the show.

4. Be sure to take power cables and coaxial cables. Throw an extra one or two in your suitcase.

5. Take tools in the display box (plus enough in your suitcase to open the display box).

6. Take tape, poster paper, and markers.

7. Arrange your literature in a manner which makes it easy to retrieve any sheet quickly.

8. Arrange the hospitality suite, too. Provide refreshments, and possibly some technical material. This is a good place to qualify prospective customers.

9. Preprint cards, inviting customers to the suite, leaving the suite number blank to be filled in at the last minute.

10. Put a sign on the elevator and one on the door of the suite to welcome customers.

11. Distribute a duty roster for the booth (don't fill it with your people; fill it with customers!) and for the suite.

12. Are any of your employees giving papers at the conference? Have a large number of reprints to hand out. Make sure the speaker is at your booth to answer questions for the next couple of hours after the paper is given.

13. Have plenty of Literature Request cards, to be certain any visitor can get more information by filling out a card.

14. Meet at the booth an hour before the area is open on the first day and show all the salespeople what is there. Give them tips on any idiosyncrasies of the demonstration equipment.

15. Cooperate with the convention staff. Attend banquets and/or other general activities, observe official booth hours, and support the activities of the sponsoring agency or group.

The week of a technical conference is a tiring, frustrating experience, but it's the best chance of the year to find new customers. Make the most of it.

DEVELOPING THIS SUBJECT

1. Imagine yourself transported back to when your employer introduced your company's most popular product. Prepare a new product release of 100 words or fewer to describe the product. What type of magazine would you select to publish the release?

2. You are charged with setting up a booth 20 feet wide and 10 feet deep at a trade show. Choose the products from your company's catalog that should be featured, and design a booth to feature them. Made a list of items to be considered in setting up and staffing your booth.

12

Your Technical Paper as a Sales Tool

WHY THE TECHNICAL PAPER?

T*HE WELL-CHOSEN,* well-written, well-presented technical paper is a very significant sales tool. Technical conferences used to offer the opportunity for equipment designers to present their best accomplishments to other equipment designers. Some of the specialized technological fields still offer such symposia. The typical trade show of the 1990s has an entirely different emphasis, however. The technical paper in a conference is now more likely to be a "sales pitch" than a design disclosure. That's the way it is, so join the crowd!

Reasons for Presenting a Technical Paper

There are several reasons to present a technical paper on a new product or system. First, those who hear the paper are more impressed with your company and your product than if you visited their offices and gave an obvious sales pitch (even though the material may be the same). Second, the conference proceedings are read by many others even though they do not attend the sessions. Companies have been contacted even 2 or 3 years after by someone who did a "literature search" and found the product to be of potential value. Third, reprints of the paper go to your Sales Engineers and give them opportunities to further your cause with potential customers. Finally, the individual engineer or programmer who presents a paper enhances his or her professional status.

There are also reasons not to present a paper: "It is difficult," or "It takes time," or "My product is not really very important anyway," or "Someone else in the company could do it better," or "Next year is better," or "This is the wrong type of conference," or "I don't like to stand before a group," or "I may be criticized." These are the same reasons the competitors' engineers and programmers are using, but aren't you more qualified than they are? So do it!

The conference at which the paper will be presented has certain rather simple rules, such as the maximum length and the general format. Read the rules—and read a few papers from last year's session. The latter may be the best way to see what is customary and acceptable.

Incidentally, consider the possibility of using a coauthor. Another person in your company might be the ideal one to make it an engineer/programmer paper, an engineer/technician paper, or an engineer/supervisor paper. Another possible coauthor is a customer. The customer contributes something about the application, and the name and title add credibility to the paper, letting the world know that you have really sold your product. You may do 90% of the work, but don't let that bother you. The result will be worth your effort.

The typical format will contain:

1. Title (brief, descriptive, suitable for the literature searcher),
2. Abstract, probably 50-100 words (simple, easily readable) to convey the gist of your paper to the casual reader.
3. Introduction (to tell the reader why your paper may be of value to him).

4. Body (with a good, easy-flowing outline, and with liberal use of subtitles to make it readable and re-readable).

5. Conclusion (to emphasize your strong points again).

6. References.

7. Appendix (only if you need it).

8. Illustrations (not too much detail on each one). Show an equipment photograph if one is available.

WRITING YOUR TECHNICAL PAPER

To write the paper, start anywhere. Too often we get bogged down with concern about what the first words should be—so why not save them for last? No one but you will know, and you'll be better able to choose a really effective statement after the entire paper is finished.

Find a critic. Avoid the person who says, "Oh, that's great." That is not helpful criticism. Find someone who does not have prior detailed knowledge of the subject to help you see how best to organize your material for the educated, but not specialized, reader. Then find a good technical writer if you want the paper "fine-tuned." You can learn a lot by watching the experienced, adept writer at work, and the writer can learn a lot about the subject by helping you with the paper. Finally, allow yourself plenty of time. That's the only way to get a good product!

Incentive payments for those who prepare and present technical papers are important. Institute a program within your company to sponsor such activity, and to reward your creative writers/speakers.

Publishing your technical paper as a magazine article is also appropriate. Look into the possibility of such publicity. True, it is a bit more difficult to get an article accepted, and it must be more tutorial than the technical paper, but success in publishing an article enhances your company's position in the high-technology field.

Example 12-1. WACKY's management had recognized in the past the value of technical papers and had established an award for each employee who presented a paper at a national conference. One young programmer, motivated by the award and a trip to a conference, submitted an abstract on an interesting subject. The abstract was accepted by the program chairman, and the programmer was given 4 months to submit the paper.

Unfortunately, the programmer didn't like to write, and found a score of excuses for postponing the task of writing the paper. Twenty-four hours from the deadline the programmer hadn't a word to start the page. The programmer then stayed up all night to write a paper that was of unacceptable quality and proved embarrassing to the company.

Remember: Either plan ahead, or don't offer a paper.

PRESENTING YOUR PAPER

When you present your technical paper to a meeting, your audience will be literate. Treat them as such! Don't insult them by reading the subject matter word for word! The paper is not written in conversational style, anyway, and will bore most of the audience quite rapidly. So put some life into the presentation.

Take some time to liven up your presentation. Lead the listeners through some of the background. Relate a personal experience if it helps to amplify the subject. Humor is not likely to be appropriate, so forget it, but make liberal use of the personal aspect of the subject. Think of scenarios in which the product or system is appropriate, and throw in one or two. Give some examples of the difficulties you encountered before you introduced this product. There's a lot to tell, so feel free to tell it.

You will have time restraints. Find out several days in advance what your session chairman has allocated for you, then plan around it. Chances are you will talk faster at the presentation than usual and will need to slow down, or plan for a fast presentation and throw in one or two more examples. Write some mileposts into your notes: "5 minutes to go," "3 minutes to go." Make certain you save time for your closing argument.

You will be excited, worried, uneasy—all those things, and more. Everyone is. One of the best ways to minimize the effect, is a lot of rehearsal in front of one or more of your interested cohorts. You can work out the kinks and build up your confidence this way.

VISUAL AIDS IN TECHNICAL PRESENTATIONS

Your visual aids deserve even more attention than your manuscript, so plan your visual aids to enhance the value of your paper. The perceived value of your paper and the image of the company will be affected by the visual quality of your presentations. Good slides can do much to enhance the oral presentation. Poor visual aids will ruin even the best written report.

The guidelines provided here, based on those of the Instrument Society of America (ISA), will help you to plan your slides, judge their legibility, and prepare for their use.

No other format is as effective as 35-mm slides for presenting visual materials to large audiences. They are easy to handle, not fragile, simple to use, and projection equipment is universally available. The conference usually provides a 35-mm slide projector, slide change signal device or remote slide control, pointer, and a projectionist.

Chalkboards are not suitable for large audiences, so they aren't normally available. Overhead transparencies, because of their size, tempt many speakers to put too much information on each frame. Overhead projectors are available at some meetings only by special request.

One note on traveling to the symposium. Make certain, after you take all the trouble to prepare your slides, that nothing happens to them. Do not entrust them to an associate, an airline, or any other shipper. Carry them with you. It's the only way to guarantee that they'll be where you need them.

Color slides are most effective for presenting photographic and line images. They cannot carry a lot of detail or complex information. To be effective your slides must be easily read and understood. If they're hard to read or understand, you'll lose your audience and your work will be wasted.

Presentation

Slides accompanying technical presentations have only one purpose, that is, to provide clarity and emphasis to the author's presentation. Confusing and distracting slides are far worse than none at all. How do you ensure clarity?

❧ Keep things simple.

❧ Use only one main idea on each slide.

❧ Use a slide series for progressive information.

❧ Avoid complex graphs.

❧ Avoid complex tables. Use simple graphs instead. Tabular material makes a dull visual, no matter how interesting the topic.

❧ Minimize printed material to a few words, all in capital letters.

❧ Do not use the visuals as memory-joggers for yourself. Do not read them to the audience. Point to an item and explain it, but let the audience read it.

❧ Limit the use of mathematical equations.

❧ Make certain your slides are legible from whatever distance you expect your audience to be seated.

❧ Put a title on each slide.

A word about specialized material and audiences. It is tempting to think that an audience of technical people need, want, and can absorb a lot of detail on slides. But nobody can handle an information overload. Complex formulas, engineering drawings, tables of data, and the like should be in your manuscript and need not be repeated during your verbal discussion. If a reference is absolutely necessary, show a simplified version of the original. Another alternative is to use several slides which break out the visual elements of the complex idea.

Color slides are more effective than black and white, simply because colors go far in maintaining audience interest. Some points to remember about color:

☑ Use colors for diagrams and charts; bright colors are especially effective.

☑ Backgrounds should never be completely white. This reduces legibility and creates a distracting glare.

☑ Make certain the use of the color is functional, otherwise it is merely decorative and not useful for your purposes.

Rehearse your visual presentation by projecting your slides well in advance of the presentation date. Examine each one for at least 3 minutes. Look for things that might confuse your audience or stimulate questions you may find difficult to answer.

Be sure the slide can be read by a large audience. A simple test is to view the slide at ten times its major dimension. If the slide is not satisfactory, have another made. The extra time and cost are worthwhile to improve the clarity of your presentation.

Avoid returning to a slide by having extra copies interspersed where you need them. Going back through the slide magazine is not a good practice, as it is both distracting and time wasting. This also may seem obvious, but do not project a slide you do not plan to discuss.

After you decide on the sequence for your slides, number the slides on the lower left corner so that they read correctly when viewed. The projectionist will rotate the slide 180 degrees for correct placement. Mark your notes with the slide numbers at the proper place.

Give your slides to the projectionist before the session begins. If you have a large number of slides, you may find it convenient to carry them already loaded in order in a slide tray. The projectionist will have empty trays if you need them. Allow yourself time to discuss your needs with the projectionist, including signals to advance to the next slide. If the projector is equipped with a remote slide control, and you become the projectionist, try it out before your scheduled session.

Recover your slides immediately at the end of your session. Don't allow eager questioners to prevent you from recovering your slides. Many meeting administrators have extensive collections of unclaimed slides. (Marking your slides and their container will help to identify the owner.)

Preparation

Unless you are a professional artist, or a very talented amateur, you should have a graphics expert produce the artwork for your slides. Your presentation and your slides will reflect on you and on the company, so planning ahead and preparing professional slides can have a high payback as good public relations.

The universally acceptable standard is the 2" x 2" (50 mm x 50 mm) 35 mm mounted slide. The image of a 35-mm slide is about 1 ½" wide by 1" high; you must keep your copy in this 3:2 ratio. To ensure clarity, original artwork for slides should be prepared within overall dimensions of 6" x 9". To test for legibility, look at the slide art from a distance 10 times the longest dimension (that is, 90 inches or 7 $\frac{1}{2}$ feet).

Always use a horizontal format because most meeting room screens are wider than they are high. If the projector is moved forward to get the full vertical on the screen, the horizontal slides' legibility will suffer. If it is not moved, the effectiveness of the vertical will be lost because only part of it will be on the screen.

For slides containing only text, a good rule of thumb is to limit the text to no more than 100 characters. It is not necessary to fill the slide. The fewer words and characters there are, the more effective the slide's visual impact.

If you use overhead transparencies instead of slides, this guide may help to make the presentation as effective as possible. Overhead projectors are available at most meetings by special request.

1. Format—horizontal.

2. Lettering—no more than 50 spaces wide and no letter smaller than $\frac{1}{40}$th of the height of the transparency.

3. Reverse type (white type on a black background) is often effective for presenting line art or lettered material because the projected images show up reasonably well in dimly lighted rooms. The lettering can be color coded to separate topics for the viewer.

4. The overhead transparency projector has the capacity for sequential overlays. Here are some points to keep in mind if you use this technique:

   ◆ Remember the principles discussed in our section on subject matter, and limit each sequential overlay to one idea. Each overlay increases the complexity of the image on the screen, so four or five add-on components are the absolute maximum.

   ◆ Captions—avoid clutter. Plan your captions so the final image contains no more than 8-10 words.

   ◆ Avoid the temptation to hand letter on overhead transparencies. If you are going to do some work at the presentation, limit yourself to arrows, circles, and the like. Simply pointing is the optimal technique.

DEVELOPING THIS SUBJECT

1. Choose one of your company's most innovative products. Prepare the title, abstract, and outline for a technical paper describing this product to be presented at a convention.

2. Prepare three or more sketches from which artwork could be made for slides on the product above. Make one or more pictorial sketches and at least one text-only visual aid.

Sample Proposal

SECTION 1: INTRODUCTION

We at the Data Acquisition Company, Inc. (DACCO) are pleased to present this Technical Proposal to the Interplanetary Rocket Research Agency (IRRA). As described in the following sections, we propose to deliver a state-of-the-art Data Acquisition System in accordance with IRRA Request for Proposal Number IR99636-E-89 dated _____ (with amendments _____ and _____).

Set the stage by telling them who you are (your official name and your familiar name), and on what basis your Technical Proposal is submitted. Start here to do the selling; describe your proposed product as a "a high-speed system," a "state-of-the-art box," etc.

We have analyzed your requirements for this system, to be used on your Rocket Test Stand Program, and the turnkey system we propose is ideally qualified to meet today's requirements. In addition, through use of our new _____ as a front end, our _____ high-speed preprocessing, the _____ computer, and our _____ software, this system has the capability for expansion in several ways. System throughput rates are more than three times your present requirement. The system can support double your specified input rates. Up to _____ real-time displays can be added to your initial quantity of color-graphic terminals. Added to these expansion capabilities are the versatility and friendliness of our software system, introduced in_____ _____ and used in large-scale systems for the _____ and at four other locations in applications similar to yours at IRRA.

The customer's program is important to him. Mention it here. Start talking about the features here, especially those that qualify you in a positive way. Build up your expandability, for example, if you feel that this is meaningful (whether the customer said so or not).

Since we are one of the world's largest suppliers of data acquisition systems such as this one, we anticipate no problems in providing your system within the _____ time frame.

The necessary space, personnel, and program manager are presently available for supplying this system, since we have just finished a large system of a similar type for _____.

Don't just say you can deliver on time; clarify why you are able to do so.

We recognize the importance of this significant Rocket Test Stand Program. Because we value greatly the working relationship we have always enjoyed with IRRA, and because this important system requirement can be met and exceeded by our standard hardware and software, we hope to be awarded this contract for the system as described.

Don't just say that you have the capability; explain the circumstances that brought this to be. Don't tell any bad news here, of course, but tell all the good news.

1.1 SYSTEM FEATURES

Features of our proposed offering are shown throughout the proposal. Some of the more significant ones are listed below:

* ❖ Totally integrated hardware/software computer/display system.
* ❖ High throughput rates, made possible by use of distributed processing power.

Show features: make the items short, snappy, and quantitative wherever possible.

* ❖ Front end is new state-of-the-art, modular _____ subsystem from one of the world's leaders in _____ systems.
* ❖ Processor is state-of-the-art 32-bit super minicomputer from one of the world's leaders in _____.

❖ Unique hardware device is used to relieve computer loading by_____.

❖ Operating software is a versatile multiuser system, well known by programmers, and enhanced by high level scientific compilers.

❖ Applications software is engineer oriented, powerful, high speed, and modular for expansion.

❖ Multiple data displays are modern microprocessor controlled color graphics, with large _____-inch screens.

❖ Powerful, versatile diagnostics are supplied for _____, _____, _____, and _____.

❖ Provision is made for addition of _____, which can _____.

❖ _____ is a state-of-the-art computer-controlled device, that can be optimized for _____.

❖ Instrumentation tape recorder is the latest and most versatile machine available, with internal microprocessor.

❖ All data is time tagged automatically by input hardware. Multiple streams can carry independent time annotations.

❖ Data routing of multiplexed input data is via computer-controlled switch matrix for accuracy and speed.

❖ Independent _____provides color graphics and/or alphanumeric data display even when the host computer is not in use. Even a _____can be incorporated here at additional cost.

❖ Most setup is in high level language, using menus. Command translation for most front end equipment is inside the boxes rather than in the host computer.

❖ System is ready for full use, but modular hardware and software make it easy to enhance operation during the life of the equipment.

❖ Support includes installation, demonstration, spares recommendations, and professional training.

❖ On-site field service is available through service contract, which can include any desired level of participation.

❖ All of this provided by DACCO, of course!

1.2 PROPOSAL ORGANIZATION

To highlight the features of this System Proposal, trade-off items of special value are listed in the following manner:

PERFORMANCE TRADE-OFF

The Problem: _____

_____

_____

_____

Our Solution: _____	Rejected Alternative: _____
_____	_____
_____	_____
_____	_____
_____	_____
_____	_____

This type of performance trade-off gives you an excellent opportunity to build up your product and shoot the competitor's technology down (without mentioning the other guy's name, of course). A blocked display such as this will always stand out—so make it good.

Topics are arranged in the order of data flow. After the System Description in Section 2, the subsequent sections describe the Hardware and Software (Section 3), the Program Management (Section 4), our Specification Compliance (Section 5), and our Corporate Capabilities (Section 6). Details are provided in the Appendix, for reference.

Lead the customers through your proposal this way. They will appreciate the help.

To define growth capability in this subsystem, it is possible to:
a) Add an_____ (4.1).
b) Add _____ (4.2.10).
c) Add a second _____ with automatic switchover
for _____ (4.3).
d) Add an _____ in the _____ (4.5.5).
e) Add _____ in the _____ (4.5.5).

f) Add a second identical _____ (as 4.5), which can merge
with the first through an additional port in the _____ (5.1).

g) Add _____capability, using an _____ .

h) Add _____capability with a bank of _____

i) Add an _____to handle _____ (4.5.6.6).

j) Add more _____(4.5.7).

k) Add a _____for _____.

*The customer may find that some extra money is available
now or later. Put in a plug for a share of that money, by showing some areas for future growth in your system design. It's a
good sales pitch; even if they never get the money, they still
don't want to feel boxed in by a limited system architecture.*

SECTION 2: SYSTEM DESCRIPTION

The Data Acquisition System for the IRRA Rocket Test Stand accepts up to 1000 measurement voltages from the test stand, stores
them, processes selected measurements, and displays the results for six
data analysts as illustrated in the simplified functional diagram, Figure
2-1A. The system is capable of collecting up to one million measurements per second, and is expandable with minor hardware changes to
double that rate if needed. Judicious use of distributed processing techniques in the system, with the use of data compression and preprocessing ahead of the computer, enables us to achieve these data rates
with a relatively small central processor.

*Provide a not-too-complex overview of the system in this
section. Use a block diagram that is easy to read and understand. Leave out those details that are not germane to the
overview. In this case, the setup/control paths are not shown,
but are referred to on the drawing.*

*Do some selling here. Show growth capability if you offer it.
Use terms like "Judicious use of distributed processing techniques..." The managers will read this far, so make it interesting and meaningful to them.*

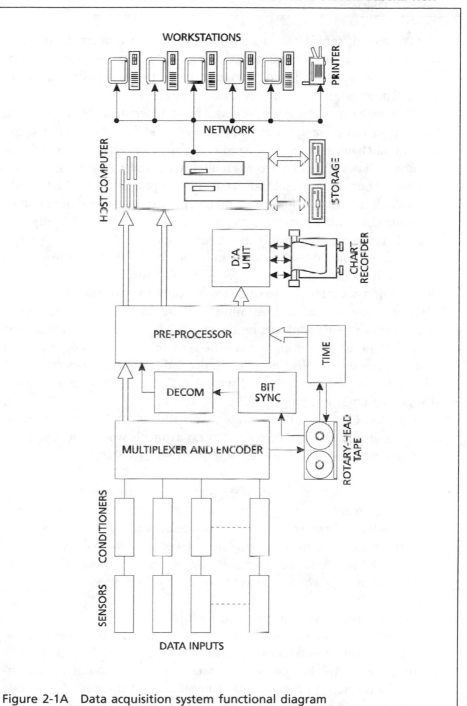

Figure 2-1A   Data acquisition system functional diagram

As shown, sensor outputs are conditioned to a normalized range in the modular signal conditioners. Personality adapters in these modules enable IRRA to use up to eight basic types of sensors and have each type conditioned properly in the system.

The high-speed multiplexer-encoder has a programmable sample plan. Each input can be sampled as frequently as desired, as long as the aggregate of all channels does not exceed one million samples per second (expandable through use of a second multplexer-encoder in the same box if needed). The encoder resolution is 12 bits (1 part in 4096). The device outputs all data in two formats: a bit-parallel word-serial real-time output is compatible with a computer or preprocessor, and a bit-serial output is suitable for storage on an instrumentation tape recorder.

The real-time output is routed to a state-of-the-art hardware preprocessor, which has the ability under program control to examine up to one million samples per second (expandable to two million by addition of modules). Based on examination by one or more algorithms, each measurement word can be discarded if it proves to be redundant or otherwise uninteresting, or can be sent to the host computer as a binary number or a floating point processed number (representing engineering units of measurement, typically). In the usual case, data rates to the computer in an application like this are only 5% to 10% of the input rates. Obviously this represents a valuable aid to the computer, and enables a supermicrocomputer to do as much work as a superminicomputer or even a minisupercomputer in this system.

If your proposed system offers a special efficiency to the user, such as enabling " a supermicrocomputer to do as much work as a superminicomputer or even a minisupercomputer...," say so. Even though the customer may have expected such performance, say it anyway. It shows them that you concur in their specified performance criteria and know how to make it happen.

The serial data stream from the multiplexer-encoder is in an industry standard pulse code-modulation format, and can be stored on the rotary-head instrumentation tape recorder in the system. This device can archive the 12 megabit per second data stream for 4 hours on a standard tape cartridge (2 hours if the data rate is ever doubled). Time code is archived on a tape-edge track for later use in data analysis.

Industry standard formats from tape playback are synchronized and converted into a bit-parallel word-serial data stream identical to that output in realtime, and are input to the hardware preprocessor. Playback time or real-time is always available for preprocessor input along with the data.

There are two paths from the preprocessor to the host computer. One is for servicing displays, and the other for data to be stored on disk. Display data builds and maintains a current-value table (CVT) in computer memory, where each measurement point is assigned a unique memory location. The current-value is always present at that location for access as a global common area in memory, to be accessed by display software. Data for disk storage builds buffers in computer memory; as each buffer is filled, computer software sends the data to a disk and the other buffer is started. These ping pong buffers transfer up to 400,000 bytes per second.

The general-purpose computer has a bus-oriented architecture. That is, data can be input, output, or moved internally with minimum participation by the processor itself.

Two Winchester-technology data disks archive information from the test stand. Six microprocessor-powered display workstations present processed data to system analysts, and a laser printer affords data logging capability. Not shown are the internal housekeeping peripherals, primarily two program disks and a streamer tape.

Raw or preprocessed data from the hardware preprocessor can be routed to a DAC/discrete unit. There selected channels are chosen and either converted to analog or output as discrete bits. A strip chart recorder displays data from this device.

System software provides all the control for this Data Acquisition System. The computer under software control can set up and control the front end devices, set up and control the preprocessor, process and display selected data channels, archive selected information to disk, and play back data from the input instrumentation tape or from disk for a second look.

The following section gives details on both hardware and software functions.

If you are hardware oriented, you may shortchange the discussion of software, or vice versa. With help from your colleagues in the other technical disciplines, if necessary, one person must write this entire section.

The Data Acquisition System is housed in three standard equipment racks, one on wheels for movement to the Rocket Test Stand instrument room. The roll-around equipment operates at 0°C to 50°C, and the other equipment operates at 10°C to 40°C. All equipment uses 115 volt ±10% 60 Hz commercial power.

It is easy to overlook the important details like operating environment. Be sure to address them briefly in this overview.

SECTION 3: EQUIPMENT DESCRIPTIONS

3.1 SYSTEM HARDWARE

Table 3.1   Hardware devices used
in the Data Acquisition System

QUANTITY	DEVICE

The hardware devices used in the proposed Data Acquisition System are listed above in Table 3.1, and described on the following pages.

The list assures the customer that you understand the scope of the job. You also need it for estimating cost,. It will be a big help to your program manager and contract administrator.

3.1.1 PCM Bit Synchronizer

The proposed bit synchronizer is a proven product operating to 5 megabits per second.

This brief (no more than two typewritten lines) highlight, probably taken directly from your storyboard, should make a specific and meaningful introduction to the subject matter discussed more fully in the next subsection.

The DACCO 720 Pulse Code-modulation (PCM) Bit Synchronizer is described in this section. Figure 3-1 shows the front panel.

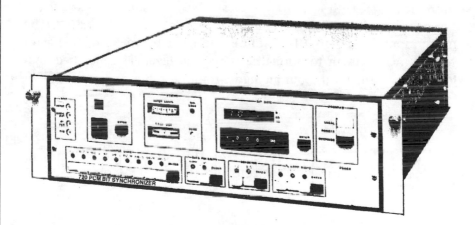

Figure 3-1 DACCO 720 PCM Bit Synchronizer front view

3.1.1 Features

The PCM Bit Synchronizer was selected for this Data Acquisition System for the following features:

★ Bit rates tunable from 1 bps to 5 Mbps for all codes.
★ Unique matched filter provides optimum performance in both filter sample and reset integrator modes of operation.

★ Exclusive bit decision feedback.

★ Performance within 1 dB of theoretical.

★ Computer and front panel programming capabilities in one model.

★ Double memory feature enables retention of front panel or remote programming setup.

★ Override feature permits front panel modification of any parameter programmed by computer entry.

★ Pushbutton and thumbwheel switches for ease of operation.

You can be certain that a list of briefly stated features will be read, so make it meaningful. If you waste the customer's time on the first few items, the balance will be ignored.

3.1.1.2 DESCRIPTION

Figure 3-2 is a functional block diagram of the PCM Bit Synchronizer. The first signal processing function of this unit is to reconstruct the PCM signal with a minimum of symbol errors and to extract the timing information that is required for further digital processing. The noisy input PCM signal is amplified and fed to an automatic gain control (AGC) circuit to normalize the signal level. The matched analog filter maximizes the output signal-to-noise ratio.

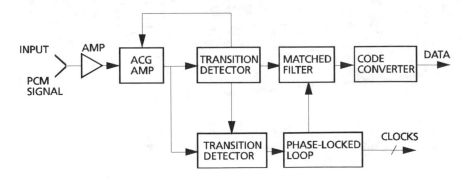

Figure 3-2 PCM bit synchronizer functional block diagram

Next, timing information is then extracted from the PCM signal by the transition detector. The phase lock loop in turn generates a set of clocks related to the transitions. One of the transition clocks is used to sample the output of the matched filter at the optimum time to make a bit decision. The symbol decision circuitry decides whether a symbol represents a logic one or a logic zero. Once the decision is made, the symbol output is code converted to produce the desired PCM output.

As you might guess, this is a very complex unit. Make its operational functions look simple in the proposal, however, to guarantee that they will be understood. Don't lose your reader in a tangle of complex drawings and words.

Table 3.2 Signal inputs of the PCM Bit Synchronizer

ITEM	DESCRIPTION
Source	Up to four input sources.
Source Isolation	Isolation between channels is better than 40 dB with 100 ohm source impedances.
Configuration	Single-ended and referenced to signal ground.
Polarity	Signals of either polarity or signals symetrical about signal ground.
Level	From 0.5 V to 30 V peak-to-peak without adjustment.
Impedance	For all bit rates, 10k ohms or greater shunted by less than 150 pF.
Maximum Input Voltages	Up to ±60 V peak when not terminated at the input.
Codes	NRZ-L, NRZ-M, NRZ-S, Bi-Phase-L, Bi-Phase-M, Bi-Phase-S, DM-M, DM-S, and RZ.
Tunable Bit Rate	Tunable from 1 bps to 5.0 Mbps for all codes.
Tuning Resolution	±0.5% of any bit rate by means of front panel or remote control.
DC Offset	Accepts single-ended signals of either polarity with the following DC offsets: 1) For bit rates > 100 bps, any value for which the peak signal plus offset does not exceed 60 V. 2) For bit rates < 100 bps, 50% of the peak-to-peak signal amplitude.
Baseline Shift	No degradation in performance if the input serial PCM wavetrain is shifted by a superimposed sinusoidal waveform with a peak-to-peak amplitude equal to the peak-to-peak amplitude of the PCM wavetrain and with a frequency up to 0.1% of the data bit rate.

This bit synchronizer is a very flexible unit. It can accommodate a number of different PCM codes, and can be optimized for performance in a variety of noise environments.

Tables, charts, and illustrations command attention. Narrative sequences do not. Use tables freely where there is a significant message to be conveyed. Use graphs (next page) where they will help to get a message across.

Table 3.3 Synchronization of the Bit Synchronizer

ITEM	DESCRIPTION
Acquisition Time	Less than 100 bit periods for an incoming bit rate within ±2% of the selected bit rate.
Capture and Tracking Range	For bit rates of 10 bps and above, either 1%, 3%, or 10% of selected bit rate with respective loop bandwidth of 0.1%, 0.3%, and 1%. For bit rates below 10 bps, ranges are fixed at 10% with a 1% loop bandwidth.
Standard Loop Bandwidth	Ranges of 0.1%, 0.3%, and 1% of nominal bit rates may be selected. Loop bandwidths up to 5% can be accommodated.
Loop Stability	Internal clock frequency is maintained to within ±0.5% of the selected bit rate with input grounded, and to within ±1% of the selected bit rate with white noise not exceeding the maximum input signal level specifications.
Biphase Ambiguity	For a biphase format having 200 continuous random bits and 2000 constant logic ones or logic zeros and operating with a SNR of 9 dB, the unit will resolve ambiguity within the 200 random bits and hold phase over the 2000 constant logic ones and zeros.
Minimum Transition Density	Sync is maintained with transition densities as low as one transition in 64 bits with a SNR of 10 dB.
Synchronization Threshold	Minimum SNR for sync acquisition is 0 dB. See Figure 3-4 for more definition.

Figure 3-3 shows the acquisition time compared with input bit rate deviation, and Figure 3-4 shows the synchronization threshold compared with transition density.

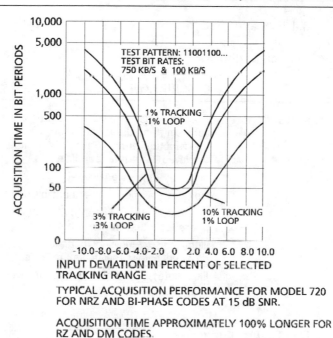

TYPICAL ACQUISITION PERFORMANCE FOR MODEL 720
FOR NRZ AND BI-PHASE CODES AT 15 dB SNR.

ACQUISITION TIME APPROXIMATELY 100% LONGER FOR
RZ AND DM CODES.

Figure 3-3 Acquisition time compared with input bit rate deviation

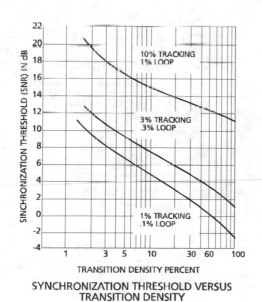

SYNCHRONIZATION THRESHOLD VERSUS
TRANSITION DENSITY

Figure 3-4 Synchronization threshold compared with transition density

Table 3.4 lists and describes the front panel displays.

Table 3.4 Front panel displays

ITEM	DESCRIPTION
Functional Status	The contents of all static stores are displayed by light-emitting diode indicators under the appropriate function titles. Bit rate and input source are displayed by numeric indicators.
Input Meter	Indicates input signal level.
Deviation Meter	Indicates deviation of the input bit rate from nominal.
Sync	Illuminates to indicate that the internal clock is phase-locked with the input signal.
Signal Loss	Illuminates to indicate that the input signal level has dropped below the minimum acceptable voltage.
Power	Illuminates to indicate that primary power was applied to the unit.

3.1.2 Digital/Analog Converter

The proposed Digital/Analog Converter is expandable to handle up to 256 channels.

The DACCO 8350 Digital/Analog Converter (DAC) is described in this section. The front panel is illustrated in Figure 3-5.

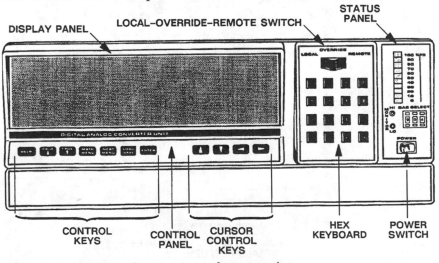

Figure 3-5 Digital/Analog Converter front panel

The control panel of your product is your customer's inter-face to the device. Show this panel prominently. If the device is yet to be developed, show your proposed front panel and mark it "proposed."

3.1.2.1 FEATURES

This Digital/Analog Converter was selected for the Data Acquisition System for these features:

☑ Supports up to 256 analog/discrete outputs.

☑ Compatible with many sources that output parallel data and timing signals.

☑ Accepts tagged data in serial or parallel format.

☑ Calibration capability with front panel controls.

Show the meaningful features in tabular form. The reader may overlook narrative, but will almost certainly read the tabulated items.

3.1.2.2 DESCRIPTION

The DACCO 8350 complements the DACCO digital product line by providing the means to convert either raw or processed data into an analog form suitable for examination on either an analog display or chart recorder. An optional feature permits the user to strip out digital events as discrete outputs for display purposes. A block diagram showing the general interface of the circuit cards in this unit is provided in Figure 3-6.

Be careful not to show too much detail on the first block diagram for each device. Keep it simple for a basic introduction; show or describe details later.

The DACCO 8350 comes equipped with a local control front panel and an intelligent interface (INF) card. This allows complete control of the unit from its own front panel, which includes initiating a calibration sequence. This version accepts data and tag in either parallel or serial format. The Word Selector will permit the input of parallel data words with associated word and frame rate timing signals as might be output from most digital decommutators.

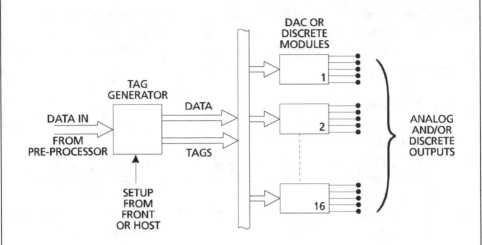

Figure 3-6 Functional block diagram of Digital/Analog Converter

The unit has seven card slots assigned to options. If selected, the Word Selector consumes one slot, leaving six for DACs and discretes. Each DAC card has 16 analog outputs and each discrete card has 64 discrete bit outputs. However, a Model 8350 has a maximum of 256 outputs. Therefore, it can support:

- Four discrete cards (no DAC).
- Seven DAC cards (without Word Selector).
- Any combination of these cards, not to exceed 256 outputs.

A calibration sequence can be initiated from either the front panel or via the host serial interface. The selected sequence may be any one of:

- Seven steps.
- Thirteen steps.
- Ramp of 4096 points.

Dwell time of the sequence steps is programmable at 0.5, 1, 2, or 4 seconds per step. The sequence can be either a single cycle or a continuous cycle. The user also can assert a fixed binary pattern into any one or all DACs.

A Word Selector is optionally available to allow interface with devices that provide data in bit-parallel, word-serial form with associated word and frame timing signals. The Word Selector performs the following functions:

• Translates word and frame counts into unique identification tags or addresses for DAC and Discrete cards.

• Converts data in sign magnitude, two's complement, or one's complement notation to offset binary form.

• Outputs data/address in serial form for use elsewhere.

Once again tabulate the functional overview. The customer will read these items.

Table 3.5   Specifications

ITEM	DESCRIPTION
Parallel Inputs	32 parallel data lines, 16-bit tag and strobe
Serial Inputs	Differential line receivers for 48-bit serial input of 32-bit data and 16-bit tag. Data, clock and strobe lines provided.
Outputs	Maximum of 256 DAC and Discrete outputs combined.
Configuration	Eight card slots assigned as:
	Card Slot 1: Intelligent Interface (INF)
	Card Slots 2 through 4: DACs only
	Card Slots 5 through 7: DACs or Discretes
	Card Slot 8: DAC, Discrete or Word Selector
Retag Memory	16K words x 8 bits. Allows user to associate the incoming tag value with a unique DAC/Discrete

The intelligent interface (INF) assembly controls communications between an external host computer (or the front panel user) and the internal functions. Host control is via a serial RS-232 port. The INF will, under appropriate command:

☑ Provide a menu.
☑ Provide health information.
☑ Perform setup.
☑ Recall stored formats.
☑ Perform a diagnostic.

☑ Provide "HELP" information.
☑ Perform all "MODE" operations.

The DACCO 8350 has a built-in self-test feature using the intelligent interface assembly to perform diagnostic tests. The INF generates and applies a test format, selects various signal nodes throughout the 8350 in a hierarchical sequence, and tests these nodes using signature analysis techniques. The signature analysis method creates a unique pattern or signature in response to the time-dependent behavior of the signal output of the function being tested. Required for each signature is: the signal output or the test node; a clock synchronous with the stable output of the test node; and a gating signal or window synchronous with the other two signals, which are repeatable and predictable. The test node signal, clock, and window are routed onto their respective bus lines and sent to the signature analysis accumulator located on the INF.

Each front panel contains a self-scan, gas plasma display of six lines by forty characters per line. Each character is formed on a five-by-seven dot matrix. In addition, a hexadecimal key pad and eleven function keys are provided for manual data entry. The front panel keyboard functions are scanned and the display is driven by a controller that communicates with the intelligent interface assembly mounted in the function chassis.

A "Monitor" function is located at the right hand side of the front panel. It contains the "DAC SEL" switches, the LED display meter, and the test points for monitoring a selected DAC output.

3.1.3 Module Specifications

Table 3.6 lists and describes the specifications of the Word Selector Module.

Table 3.7 lists and describes the specifications of the D/A Converter Module.

Table 3.8 lists and describes the specifications of the Discrete Output Module.

Table 3.6 Word Selector Module specifications

ITEM	DESCRIPTION
Bits/Word	Up to 16
Word Alignment	MSB
Bits/ID	Up to 8 bits (256 Unique Addresses)
Words/Frame	Up to 1023
Frames/Subframe	Up to 1023
Code Conversion	Selectable on a word basis. Performs conversion from a sign magnitude, two's complement, or one's complement to offset binary on MSB aligned data up to 12 bits in length.
Calibration	12-bit word and 8-bit address provided
Inputs:	
Data	16 bits
Word Increment	1 line
Word Reset	1 line
Frame Increment	2 lines
Frame Reset	2 lines
Data Ready	1 line
Acknowledge	1 line (output)
Serial Output	Serial form of 16-bit data and 8-bit address for remote Units. Operates with normal or calibration modes. Capability to drive three separate differential lines.
Input Levels	TTL Compatible

Table 3.7 D/A Converter Module specifications

ITEM	DESCRIPTION
Inputs:	
Data	12 bits
Tag	8 bits
Strobe	1 line
Input Levels	TTL Compatible
Output	16 analog outputs
Output Ranges	0 volts to +10 volts (-100 Version)
	From -5 to +5 volts (-101 Version)
	From -10 to +10 volts (-102 Version)
Output Configuration	Single-ended, referenced to ground
Accuracy	±0.1% of full scale
Output Current	10 milliamperes maximum
Load Capacitance	0.01 microfarad maximum

Table 3.8 Discrete Output Module specifications

ITEM	DESCRIPTION
Inputs:	
Data	32 bits
Tag	8 bits
Strobe	1 line
Input Levels	TTL Compatible
Outputs	64
Output Levels	TTL Compatible
Output Current	Sink 24 milliamperes at logic zero. Source 15 milliamperes at logic one.
Mapping	Any bit or bits from input word to any of the 64 stores

3.2 SYSTEM SOFTWARE

The proposed software has been proven in more than fifty systems and is provided as a turnkey system.

The application software consists of proven standard software products plus software developed specifically for this Data Acquisition System. More than 95% of the software is coded in VAX-11 FORTRAN and conforms to standards for coding and documentation. A small portion of the software is coded in VAX 11 MACRO to meet timing constraints and/or VMS operating system requirements (for example, device drivers, high speed I/O routines). The system is supplied with the software fully integrated and ready for operation.

You are writing this for both the hardware personnel and the programmers. Give an overview of the significant characteristics, in easy to understand terms.

The software is designed to provide a simple "user friendly" interface to programming and operation of the computer and front end equipment. A main menu gives the operator a central point of control for the operation of the entire system. From this main menu, the operator may select lower submenus that control particular subsystems. The following features, common to all our software, provide this friendly, simple interface:

❖ All operator entries are through menus with meaningful prompts and descriptive error messages.

❖ Menu entries are verified upon input, and cross-checked later against other entries where possible.

❖ Menu entries and operator commands may be stored on disk.

❖ Hard copies may be generated for permanent storage or operator reference.

❖ The use of function keys and arrow keys is consistent across the entire menu system.

❖ Defaults can be built into menus requiring many entries. This eases the creation of databases required for new formats.

❖ An entire format may be entered ("canned") once by the "experienced" user and executed later by the "casual" user. This ensures ease of operation by personnel less familiar with these systems.

As with hardware, give the important features in an easy-to-read form. Be sure again to use terms that the hardware reviewer (possibly the boss, the decision maker) can appreciate.

The system software is made up of five functional subsystems:

★ Database setup software.
★ Data acquisition and storage software.
★ Real-time and playback display software.
★ Print software .
★ Software for playback of disk data.

This is a lead-in to the detailed description. Make it meaningful, and use simple terminology.

These subsystems are discussed in the following paragraphs.

3.2.1 Database Setup

All stream characteristics and all measurement parameters are defined in the data base, and need never be re-entered.

Database setup consists of entering information required to describe the format of the data, and processing to be performed on the data. This setup takes place through interactive menus. Setup information may be stored on disk for later use. Thus, once a particular scenario is defined, it may be immediately recalled in preparation for a test. Rapid switching is possible between tests with minimal operator effort.

Setup information consists of the following:

■ A parameter database containing the information required to process a particular parameter's data within the format. This information is used to program the preprocessor, and for playback from disk.

■ A set of front end setup menus.

■ A set of DAC setup menus for selecting data to be sent to the stripchart recorder. Programmable features of the DAC modules are a part of these menus.

■ Display page definitions for real-time and playback displays.

■ Print page definitions for the tabular print mode.

■ Other information required for processing data.

3.2.2 Data Acquisition and Storage

Data input at high rates is automatic under software control.

Data acquisition and storage functions provided by the main menu system consist of loading the front end, starting data acquisition, and stopping data acquisition. The following types of data may be acquired and/or recorded by the system:

☑ Data and IRIG time from "live" tests, analog tape playback, or simulators.

☑ Data that has been digitized by the A/D card and transferred to the computer for storage on disk.

☑ Status from an analog tape.

Data enters the computer under operator control from the preprocessor via two interfaces.

EU converted data and IRIG time from the preprocessor enter the computer via the input channel, and are placed in a current-value table (CVT) for use by the real-time display software as shown in Figure 3-7. The CVT data is double buffered, so that data from a single frame is updated in the CVT simultaneously.

Simultaneously, raw untagged data from the preprocessor enters the computer via the interface. Buffers are built in memory and then written to disk for later use by the playback software shown in Figure 3-8.

Showing sketches of software functions will be difficult at first, but the technique can be developed and can prove very valuable in creating an understandable proposal. Again, keep those hardware personnel in mind!

3.2.3 Data Display Software

Graphic and numerical displays are supported in real-time and playback modes.

System software provides for the interactive display of converted and raw data during real-time (live or analog tape playback) data acquisition and during playback from disk. Features common to the real-time and playback display modes include the following:

❖ Both graphic and alphanumeric displays are provided.

❖ Operator-named setup pages may be defined and stored to disk for later recall. The number of defined pages is limited only by the amount of available disk space.

❖ Screen hardcopy (alphanumeric and/or graphic) may be initiated by a single keystroke.

❖ Time is displayed along with the data.

❖ An operator-entered 60-character title may be displayed with the data.

❖ Out of limits parameters are highlighted.

❖ Parameter names and other setups are verified upon entry. The operator is advised of an improper entry through a prompt that must be acknowledged before proceeding.

When a display program is selected from the main menu, the set-up page is displayed. The operator can enter information into the page or recall a previously saved setup page. In either case, the operator may edit the page. Then, with a single keystroke, the operator may proceed to data display.

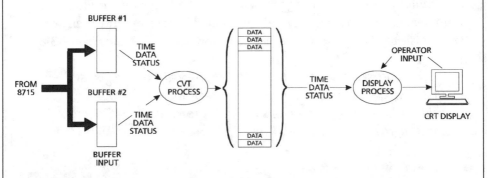

Figure 3-7 Current-value table (CVT) processing

Make it sound powerful, yet simple to comprehend. The customer will appreciate this, and may reward your efforts with a contract!

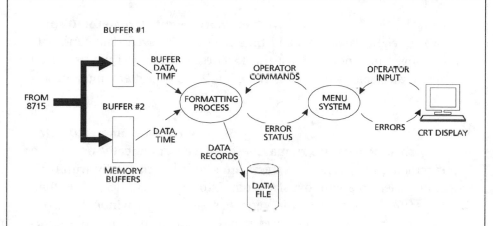

Figure 3-8 Data formatting to disk

3.2.3.1 REAL-TIME DISPLAYS

Real-time displays for the system consist of the following DACCO standard products or modified standard products:

☑ Real-time alphanumeric eight-parameter display.
☑ Real-time graphic four-parameter display.
☑ Real-time alphanumeric thirty-two-parameter display.
☑ Real-time sixteen parameter barchart display.
☑ Real-time alphanumeric sixteen-parameter alarm display.

The real-time displays have operator selectable update rates and status indicators also.

The real-time display provides a continuous scrolling capability. The entire image on the screen is scrolled horizontally. The horizontally scrolling capability of the terminal is limited: the image scrolls on 16-pixel boundaries. The smoothest horizontal scroll possible produces a perceptible jump of 16 pixels with each scroll command sent to the terminal. This jump is only 2 percent of the width of the screen image, however, and is not objectionable.

If you have a characteristic that may appear negative to the customer, consider a justification as part of your technical description. Don't go looking for negative features, but be realistic about the obvious ones.

Alphanumeric Eight Parameter Display

Up to eight parameters and time may be displayed. Time and data are arranged in nine columns; the 15 latest values scroll upward at an operator-selectable rate. Data may be logged to a printer simultaneously, if desired.

Graphic Four Parameter Display

Up to four parameters may be displayed graphically on one to four scrolling plots. The data traces are scrolled horizontally within an operator-selectable window size from 5 to 9999 seconds. If less than four parameters are used, the vertical scale of each window will increase proportionally. Data traces may be clipped to within the limits of each display window, or allowed to run outside the window.

Alphanumeric Thirty-two Parameter Display

Up to thirty-two parameters may be displayed simultaneously. A single value for each parameter is displayed along with time. The parameters are arranged in two columns. Time appears on the top left hand corner of the display. The parameter name and units appear next to the parameter value.

Sixteen Parameter Bar Chart Display

Up to sixteen parameters may be displayed simultaneously in bar chart format. The parameter names appear on the left. Next to each name is the current value of the parameter and a horizontal bar that corresponds to the current value of the parameter. The 0 and 100 % values for the bar are operator selectable.

Alphanumeric Sixteen Parameter Alarm Display

Up to sixteen parameters may be displayed simultaneously on the left of the page. On the right, fifteen alarm conditions are displayed in order of operator-selectable priority. The parameter name, status (low or high limit violation), and priority of each alarmed parameter are displayed. The name and value of the parameter with the highest priority alarm are displayed on the second line along with time. If more alarms exist than can be displayed, a line noting that condition is displayed.

3.2.3.2 PLAYBACK/ANALYSIS DISPLAYS

The Playback/Analysis displays for the system consist of the following standard products, modified standard products, and special software:

- Playback alphanumeric eight-parameter display.
- Playback graphic four-parameter display.
- Playback alphanumeric thirty-two-parameter display.

Additional features of the playback displays are the ability to select any time window in the recorded data file for display, and to search the file for specific data conditions (less than a value, greater than a value, equal to a value, etc.). All playback displays support operator-entered launch time (also referred to as "zero time" or "reference time").

If a characteristic is helpful to a user, get extra points for yourself by defining it in the proposal.

Alphanumeric Eight Parameter Display

Up to eight parameters and time may be displayed. Time and data are arranged in nine columns that scroll upward at an operator-selectable rate. The parameter name and units appear at the top of each column. An operator-selectable skip factor may be used to display every Nth frame of a data file rather than every frame in the file.

Up to fifteen parameters may be printed in an 80 or 132 column format as selected by the operator.

Playback Graphic Four Parameter Display

Up to four parameters may be displayed graphically on one to four plots. If less than four parameters are used, the vertical scale of each window will increase proportionally. Data is displayed in a window sized by the operator. Subsequent data may be viewed by entering a single keystroke. This allows the operator to view data sequentially in the forward direction (time increasing).

The Playback Graphic Four Parameter Display also supports cross plots (one parameter against a second parameter other than time). The data values for the first parameter are used as the y-coordinates, and data values for the second parameter as the x-coordinates. The points are joined until the entire plot is completed. The scales for the x- and y-axes are operator-selectable, as well as the time window over which the data is selected. The parameter name and plot values are displayed along the appropriate axis. Subsequent time windows of cross-plotted data may be selected by a single operator entry. When a new time window is chosen, existing traces are erased and a new plot is started.

Playback Alphanumeric Thirty-two Parameter Display

Up to thirty-two parameters may be displayed simultaneously. A single value for each parameter is displayed together with time. Parameters are arranged in two columns. Time appears on the top left hand corner.

3.2.4 Printing Software

The same data that is displayed can be printed in either of several formats.

A general purpose tabular print capability is built into the alphanumeric eight parameter displays for use in real-time and disk playback as illustrated in Figure 3-9. This tabular print capability may be used to print any data. The data may be printed in binary, octal, hexadecimal, decimal, or engineering units (five or more digits with floating decimal). Discrete data may be printed with state names (for example, ON, OFF, EMPTY, FULL).

Again, don't miss an opportunity to present a feature for your customer's consideration. Your competitor may overlook the opportunity.

3.2.5 Software for Disk Playback to DACs

Data recorded on disks can be played back into the Digital/Analog Converter unit for output.

The operator can select channels from the disk files for output in analog form from the system's Digital/Analog Converter, in the same manner that real-time or analog tape playback data is available for DAC output. Figure 3-10 illustrates this data flow.

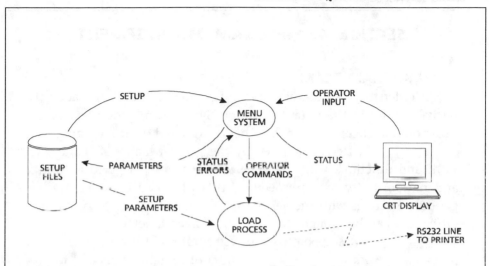

Figure 3-9 Printer setup for general-purpose tabular printing

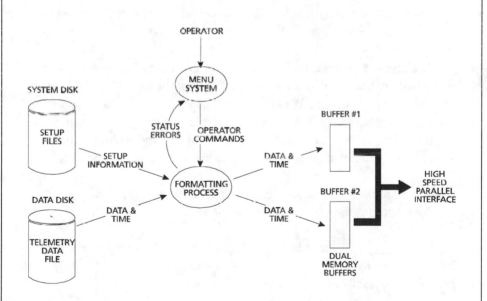

Figure 3-10 Playback of disk data to stripchart recorder

SECTION 4: PROGRAM MANAGEMENT

4.1 PROGRAM TEAM

In order to give IRRA assurance of proper technical and administrative control of the Data Acquisition System, a program team organization has been set up by the designated program manager, _____. He is appointed by the general manager of the division, and functions with the general manager's delegated authority and responsibility on all aspects of this proposed contract. He has periodic meetings with the general manager and his staff, so that any difficulties can be brought to light and solved quickly.

_____ was appointed contract administrator on the proposed program, to work with _____ in all contractual matters and to provide liaison with the contracting officer at IRRA.

Be specific here. The customer has a right to meet your program team. If you find it impossible to put the proposed people on the contract, you can "negotiate" substituting but do not deliberately mislead the customer. Promise the services of people whom you consider will be available.

We propose to hold an in-process design review every 2 months in our plant in Colorado during completion of the contract. The review will cover system design. We also propose to hold a program/production progress meeting every 3 months in our plant during fulfillment of the contract. The review will cover progress. For the months in which two reviews are scheduled, they will be combined in a single meeting.

Be specific. This is your chance to show that you have formulated a logical and meaningful plan.

_____ will provide leadership for the hardware design, assembly, and test functions as the system engineer, and _____ will provide leadership as the system programmer.

The technical people must be introduced also (with the same provision which we mentioned earlier for the program manager and contract administrator).

The subcontract with _____ for the _____ constitutes a major coordination effort. _____, subcontracting specialist, will administer the contractual details with the subcontractor through their contract administrator, and each related specialist at DACCO (engineering, quality assurance, technical writing, training, et cetera) will coordinate with their counterpart, the subcontractor through their assigned program manager, _____. _____ will monitor and control all of this interaction.

Each of these people has a counterpart at the customer organization, and their identities are important.

Purchased standard products (from _____, _____, _____, and _____, for example) will be administered through _____, with participation by others on the program team as appropriate.

All manufacturing will be under the supervision of _____, director of operations. Standard DACCO products will be built and tested in the manufacturing area; the custom items and all system packages will be assembled, wired and tested under the supervision of _____ in the system assembly area.

Don't overlook these persons.

Quality Assurance (QA) coverage of the proposal contract starts at the engineering phase, and covers hardware and software standard products, custom assemblies, and systems. The responsible quality engineer on this program is _____.

Test and inspection are monitored by QA. The test procedures will be standard on DACCO and vendor standard products; special procedures will be prepared on the system and submitted for IRRA approval. The special procedures will be prepared by the system engineers and programmers. The only device on which we propose a different technique is the _____; on this unit, we propose to have all detailed tests run at _____ because they have the unique test facilities.

Let the customer know that you have a logical, practical plan. It gives them a feeling of confidence in you. Possibly you will need to renegotiate some of these details, but lay them on the line in the proposal.

All design configurations will be monitored by the configuration management group led by _____. Any changes in design on standard products, special products, or systems will be documented for the program manager and for IRRA.

Technical manuals will be prepared under the direction of _____, with specific control of software documentation by _____, specific control of hardware documentation by _____, and specific control of subcontractor documentation as further described in paragraph _____.

Drawings will be prepared by our drafting department under the direction of _____. Level 2 drawings will be provided whenever new drawings are needed; this will include system-level designs and custom product designs. Standard products used in the system will not require production of new drawings.

Don't overlook these details of your proposed activity.

Packing and shipping will be coordinated by _____, manager of that function at DACCO.

Each person in a key position has more than 5 years of directly related experience.

Use a little salemanship here. It lets the customer know that you didn't hire these people yesterday.

In a typical year, DACCO develops and delivers 20 to 25 systems of the same relative complexity as the proposed Data Acquisition System. Some of the more recent systems we have built which are similar to the Data Acquisition System are described in the Appendix. In all cases, further details are available to IRRA evaluators on request.

Each of the systems mentioned, as well as other systems from DACCO, included some or all of the support elements proposed for IRRA:

☑ Acceptance test procedures, and detailed acceptance testing
☑ Environmental test procedures, and environmental testing at our plant or at a testing laboratory
☑ Professional quality training, both hardware and software
☑ High quality technical manuals, both hardware and software, some of them to separate specifications
☑ Field support after delivery

Our experience on these programs is IRRA's assurance of our capability to handle this proposed program with minimum difficulty, using this program team.

Use positive sales-oriented terminology here.

4.2 QUALITY ASSURANCE

4.2.1 Organization

The Director of Quality Assurance reports directly to the General Manager. This organization structure provides the necessary direct line visibility to executive management to achieve the Product Assurance Charter.

Detailed instructions on implementing the QA policy are further discussed in our QA Manual and Manufacturing Organization Instructions, each of which is available to IRRA for review. A Quality Assurance plan will be submitted to disclose to IRRA fulfillment of each paragraph of sections 3 through 7, MIL Q 9858. In summary, the Quality Assurance program utilized for IRRA will be in compliance with MIL Q 9858, MIL STD 454 Requirements 9, and established DACCO Workmanship and Quality Standards. The calibration system adheres to MIL STD 45662; our standards are traceable to the National Bureau of Standards.

Personnel in Quality Assurance who will relate to IRRA are:
Quality Engineering Manager _____
Quality Engineer _____

Quality Test   _____

Lab Analyst   _____

Electronics Analyst   _____

Reliability Manager   _____

Reliability Engineer   _____

Remember, the customer's QA team will review this section carefully. Show them that your QA organization is important and capable, and that it has authority to control quality.

4.2.2 Control of Purchased Equipment Quality

Quality engineering will review all procurement requests to assure that appropriate requirements are contained. All purchased material and equipment is subjected to incoming inspection.

On the Data Acquisition System, the major procurements of vendor standard catalog equipment are as follows:

EQUIPMENT	VENDOR	PRODUCT
Enclosures		
Receiver		
Switch Matrix		
Cassette System		
Monitor Scopes		
Graphics Terminal		
Stripchart Recorder		
Computer		
Monitor Receiver		
UPS		
Time Code		

Because each of these is a standard product from the referenced manufacturer, we propose to monitor quality at the incoming inspection check point but not to impose unique quality requirements on the manufacturers.

The quality of purchased equipment must be controlled by you also. Don't forget to say so.

4.2.3 Control of Manufactured Equipment Quality

Standard Products

All of our production is controlled by workmanship, inspection, and test standards in full conformance with MIL Q 9858. The equipment in this category includes:

Analog Recorder _____

Synchronizer _____

Preprocessor _____

In addition to normal Receiving and Fabrication Inspection, In-Process Quality Controls will assure IRRA that production processing, fabrication, assembly and test are accomplished under controlled conditions to achieve the required level of product quality. Inspection operations are identified on the Process and Routing Sheets along with the established procedures, tools, and drawings to complete required inspection.

To guard against damage from electrostatic discharge, our internal procedures comply with DOD STD 1686. Continual emphasis is placed on ESD training and awareness through supervisors and Quality Circle activities. Our Quality Assurance department audits ESD requirements at every level of manufacturing. Additional refresher training is conducted during Soldering Training/Certification Programs. Training is administered by a Quality Engineer who holds an active "Category C" classification (Instructor Examiner to MIL-S-45743E and W.S. 6536) issued by the Department of Defense.

The customer may not have called for all of this. If you provide it as part of your standard activity, tell about it anyway. It makes you look even better than they require.

A Quality Engineer reviews and approves each product acceptance test procedure, monitors in process integration (assembly), participates in final acceptance testing to assure compliance with the test procedure(s), and verifies that all test equipment is within the specified calibration dates. All test records resulting from system testing are independently reviewed by Quality Engineering and Final Quality Control department to assure that test procedure requirements have been met, and that required data have been recorded and are within specified limits.

System

The Data Acquisition System production and test are controlled by the same workmanship standards described above and based on MIL Q 9858. Acceptance tests are based on the procedures that IRRA will review and approve.

We will perform a 48-hour temperature test on the Data Acquisition System, using MIL-STD-810D Methods 501.2 and 502.2, paragraph II-3.2 as a guide at the operating temperatures of +10°C to +50°C, and will provide a summary report of results on each system.

Be sure to show the QA participation in testing. They are the customer's representatives in your plant.

Each system will receive a final inspection and verification before shipment. Packing and shipping inspection ensure that all items are adequately packaged and protected to guarantee safe arrival and identification at their final destination, and that IRRA packaging, packing, preservation, and marking requirements have been satisfied.

4.3 RELIABILITY

A reliability engineer will review engineering change control documents to ensure that the design is in no way degraded in form, fit, or function by proposed changes. Any changes that may have an adverse effect on the system will include an impact assessment from the reliability department that will accompany any ECP or other documentation sent to IRRA for approval.

Is this required by the Customer? If not, but if you do it anyway, tell about it and gain a few extra points over your competitors.

A reliability engineer will attend design reviews and provide an assessment of system reliability as well as suggestions regarding any issues related to reliability of the hardware.

A system analyst will use numbers supplied by the various subcomponent manufacturers. Where numbers are unavailable they will be estimated either by similarity with other equipment or by parts count.

There is no assurance that we can meet the MTBF stated in the IRRA specification, since most of the equipment is already defined and not subject to reevaluation by our reliability team. However, we will accept the 1000 hour goal and assess the system accordingly.

4.4 CONFIGURATION MANAGEMENT

Configuration Management (CM) procedures from our normal activities will be adapted and tailored to meet IRRA requirements defined during contract negotiations. Within 90 days of contract award, a Configuration Management Plan will be submitted. This plan will describe program specific procedures, techniques, and practices utilized in the maintenance of Configuration Accounting, Identification, and Control.

The following pre-production baseline will be established:

☑ System, plus documented pre-contract negotiated changes.

Prior to delivery of the production unit we will submit As Built Configuration Lists. The lists will identify unique hardware/documentation plus standard product and vendor items at the top assembly level.

Engineering changes, waivers, and deviations will be processed in accordance with the plan and will be submitted to IRRA.

4.5 DATA MANAGEMENT

Recognizing the importance of Contractor Data Requirements List (CDRL) items on the contract, the Data Management (DM) organization is charged with the review and submittal of all technical data in accordance with contract requirements. Throughout the period of performance of this contract, Data Management, reporting to the Program Manager, will provide administrative support to ensure that data preparation and timely submittal will be accomplished in accordance with IRRA data requirements.

The Data Manager will be responsible for review of requirements to identify and itemize contract deliverables and to develop a master schedule which includes both contractor generated and subcontractor generated data. Utilizing an automated calendar system, DM will track data items prior to submittal.

Data requirements for the subcontractor will be established in the subcontractor statement of work. All data will be monitored to ensure format and timely submittal to IRRA.

Again, possibly the customer didn't demand this. If it is standard practice, say so.

DM will be responsible for the submission of all data items to IRRA in accordance with the contractual requirements. Upon final review of the data item, the Data Manager will prepare the transmittal package. Sufficient copies of the data item will be obtained to meet the contract requirements and to provide for internal distribution and file copies. DACCO will use a standardized transmittal document as a means to provide specific CDRL identification, distribution, and/or response/acknowledgment provisions.

All data submission with the exception of released drawings will be maintained by the Data Manager. DM will track CDRL items through customer approval. Inherent in this function will be the coordination of questions, comments, and concerns between the CDRL author and the customer.

The automated calendar system provides Data Management with reports depicting CDRL items already submitted, future submission dates, and past due submissions, if any. These reports will be provided to the Program Manager to keep him abreast of the current status of data deliverables.

4.6 TECHNICAL MANUALS

DACCO will provide a complete set of technical manuals for the Data Acquisition System. The technical manuals include commercial manuals for every product included in the system, and customized system level manuals in compliance with the IRRA technical data requirements.

4.6.1 Standard Product Manuals

DACCO will deliver commercial grade manuals for all products contained in the Data Acquisition System. These manuals include DACCO standard product manuals and outside purchased equipment (OPE) Manuals:

This listing has several purposes. First, it gives your Customer the assurance that you know what manuals are required. Second, it gives you a checklist for pricing. Third, it is your Program Manager's list for the program design review, to remind the Customer that they have no reason to expect more than you promised them.

Time code generator
Switch matrix
Data encoder/decoder
Cassette tape recorder
Analog tape recorder
Monitor oscilloscopes
Terminal
Strip recorder/printer
Computer
    Tape controller and cartridge tape
    Disk controller and two disks
    Parallel interface
    Asynchronous multiplexer
Software (standard)
    Operating system
Compiler
Uninterruptable power supply

List the software (preferably in more detail than this) and such extra items as an uninterruptable power supply.

4.6.2 Special Commercial Manuals

Because the Preprocessor is such an integral part of the Data Acquisition System, and because it is more complex than any other device in the system, the technical manuals are expanded from normal to incorporate greater operation and maintenance detail.

The technical manuals provide a focal point for understanding not only the functions of the preprocessor but also many aspects of system operation and maintenance. The design of each manual addresses the fundamental needs of its intended audience, whether the reader is a novice system operator, a highly skilled maintenance technician, a telemetry site manager, or a system software/hardware engineer. The standard commercial manuals include:

- Application user's guide
- Service procedures manual
- Maintenance manual
- Distributed processing unit algorithm reference manual
- System configuration utility user's manual

In addition to these standard manuals, DACCO will provide custom system user, operator, and programmer's manuals for the system. These manuals contain modular portions with information pertinent to the remote or host user interface to the system. The most critical aspect of that interface is a set of menu driven programs that offer the user complete access to the parameter data base for command setup, load, display, and related functions. These menus are fully documented in the User's Manual which is further explained in paragraph _____.

Your software documentation should be worthy of a proud description, such as this.

4.6.3 System Hardware and Software Manuals
The Data Acquisition System publications will include:

- ☑ System Manual
- ☑ System User's Guide
- ☑ System Operator's Guide
- ☑ Programmer's Guide

The System Manual will be written in compliance with IRRA Technical Manual Contract Requirement, and will provide seven chapters as follows:

Chapter 1: General Information
Chapter 2: Safety Precautions
Chapter 3: Operation
Chapter 4: Functional Description
Chapter 5: Scheduled Maintenance
Chapter 6: Fault Isolation
Chapter 7: Installation

The overall system must be documented to enable your customer to use it properly. Propose it and explain the level of detail to be included.

The system manual will be written to DACCO style standards, which require that the manuals are prepared in double columns, with lists of effective pages, change records, lists of illustrations, tables of contents, glossaries, running heads and running feet, footnotes, engineering change notice reports, and evaluation records. Illustrations and text are prepared on a state-of-the-art computer-aided publishing system. The quality of the illustration and line drawings exceeds guidelines of the IRRA. The style standards are derived from NAVY MIL M-38784B and AF MIL-STD 7298.

The User's Guide will be specifically engineered for system users using modular techniques, and its design will correspond to the Main Menu design to ensure easy reference. The User's Guide specifically complies with the Technical Manual Contract Requirement. The intended audience is the end user of the system who performs the following functions:

- ❖ Telemetry file maintenance
- ❖ Data acquisition
- ❖ Real-time data display
- ❖ Analysis of data

The User's Manual includes help screens, error messages, a glossary, and an index.

The Operator's Manual is written for the system computer operator and contains procedures for the operation and maintenance of system files. The Operator's Guide includes the following sections:

1. Power up and boot procedures

2. Initiation of system functions
   - Operation of system functions
   - Formatting a disk
   - Initializing a disk
   - Mounting a data disk

- Creating a directory on a data disk
- Backup
- Test accounts
- Initializing a tape
- Rebuilding/configuring software
- Diagnostics

3. Error messages

4. Parameter database
   - Convert utility
   - Privileged operations
   - Copy utility
   - Derived parameter user library

5. Special operational notes
   - Disk pack special precautions (problem detection, handling, storage)
   - Magnetic tape usage

6. Graphics
   - Initialization
   - Diagnostics
   - Demonstration

The Programmer's Guide covers all special software. The intended audience is the programming staff. It provides information on the design, logic, operation, and maintenance of the telemetry system software. This manual includes the following sections:

1. Introduction
   - General overview of the software
   - System level specifications
   - Reference information

2. Technical Overview
   - Purpose/function
   - Operational description/software system design
   - System flow/hierarchical structure
   - Standard accounts and directories

3. Programming and Usage Guidelines
   • Program installation and interfacing notes
   • Operating procedures/using the software
   • Overview of available programming tools
   • Examples/typical tasks

4. Package/Library Documentation Each system is made up of major software packages and libraries. This section describes the following for each package in the system:
   • Introduction
   • Applicable Documents
   • Environment
   • System Design
     • System flow
     • Hierarchy listings
     • Module names
     • Data structures
     • Installation command files

For each library, module level titles, descriptions, and calling sequences (including definitions of arguments) are provided.

Appendices contain system messages, a glossary, and an index.

Software manuals are hard to describe. That is a good reason for you to make yours stand out in contrast to those of the competition.

4.7 TRAINING

DACCO will provide 4 weeks of in-house training. At the conclusion of the training period, user personnel will be able to install and test the system within an 8-hour period, operate the equipment with a single technician, and maintain the equipment to the card level.

Training manuals will be submitted. The primary training aids are the technical manuals on the system. The training course outline will define our lesson plan and will refer to these manuals. The approved acceptance test procedure will also be to teach operation. This has a special value to the student, since this same procedure may be used for the life of the system as a verification of hardware/software performance.

If training is part of the requirement and of your proposal, make certain that it helps sell the system. Your competition may have a good training program, too; but maybe they will not know how to describe it effectively.

4.7.1 Objective of the Training Program

The training program is designed to help users to acquire or improve their proficiency in use of our equipment. The user who already has experience in the operation or maintenance of our products or systems is likely to complete the program at a higher proficiency than one who is unfamiliar with these products or systems. However, the objective is to accommodate a given class to the needs of the students, and to give each a significant advancement in proficiency in the related subject.

Hardware training is offered at the "box" level and at the "system" level; software training is offered at the "system" level. Within those broad categories, classes are intended to help a student:

a. Operate specific hardware effectively, and/or
b. Maintain specific hardware effectively, and/or
c. Operate a specific system effectively, and/or
d. Add to or revise a specific system program effectively.

4.7.2 Prerequisites

Because the starting level for instruction in any training program must be geared to the needs of every student in the class, certain reasonable prerequisites must be established for the program. In general, these are:

a. Hardware operation: a general knowledge of _____technology, acquired through at least 3 months of experience in a _____ group (or alternatively, a 2-day familiarization class).

b. Hardware maintenance: same as above, plus completion of a technical course or equivalent on digital (or analog, for analog equipment classes) technology.

c. System operation: same as (a).

d. Programming: same as (a), plus completion of a programming fundamentals course, plus completion of a course on the specific computer operating system being used.

We do not verify conformance to these prerequisites, but expect each prospective student to do so when choosing a course of instruction. Enrollment in a class that is too advanced may result in failure to achieve one's objectives. Assistance is available in choice of a class or classes; details are available on the 2-day familiarization course that is of value as the alternative to experience in (a) above.

4.7.3 Instructors and Facilities

The Training program at DACCO is set up with a staff of full-time instructors, and also takes advantage of the in-house or local availability of specialists for part time service as instructors in their individual areas of expertise. The net result is a combination of experience and tutorial ability for each class subject.

The full-time instructors prepare training materials and provide instruction in "standard" classes, most of which are offered periodically at our plant or on special contract in customer locations. Special classes are handled by these same personnel, except when course material is so specialized as to require unique qualifications.

One dedicated classroom in the plant provides the training facility for the largest class each week. This room accommodates up to 24 students, each with comfortable chair and a shared table. An overhead projector and large screen make it possible for each student to observe all resource material on the screen and/or over-sized chalkboard.

An equipment area in the rear of the classroom provides space for hands-on operation of the related devices. Tie-in to the computer in the Data Processing Laboratory offers the opportunity to operate software without the need to move students during classes.

Even a classroom or other instruction area is worthy of special description. Don't overlook details of this type in preparing your proposal preparation.

Other classes are accommodated in smaller classrooms, each with the same type of support for optimum student observation and participation. When a class is offered at a customer location, the same type of facilities should be provided to guarantee maximum benefit for all students.

4.7.4 Visual Aids/Handouts

Each instruction plan includes liberal use of transparencies, each illustrating an element or detail for the student. Many of these illustrations are excerpted from the technical manual or operator's manual in the typical course, partly because these are representative of the material to be taught, and partly because it helps the student to become familiar with the primary resource document of any laboratory: the related manuals.

New material is developed for classes as necessary to amplify the technical documentation. Much of this material is an inevitable outgrowth of questions and discussions from previous classes on the same subject.

Student handouts include the appropriate material for the subject, and are profusely illustrated. The intent, of course, is that a student be able to mark the material as desired following instructor's comments, and then to refer back to the material with confidence months or years later.

This helps to sell your understanding of your responsibility, and helps the program manager to define the scope of activity when the contract is awarded.

4.8 FIELD SERVICE

Our Field Service organization at DACCO will provide the required service for the Data Acquisition System after delivery. This group is experienced in maintenance and operation of our equipment; both hardware and software are supported. If any problem is too difficult to be handled by the service engineer and his or her support group, it will be referred by them to the appropriate system or product expert in our company, or elsewhere.

If you have a good field service organization, it is a big selling point for your proposal. Build it up to where it belongs as a sales tool.

The assigned field service engineer will be present during factory acceptance tests at DACCO on the system, and will support the Field Operational Tests at the IRRA facility (3 weeks for the system).

4.9 SPARES LIST AND RECOMMENDATIONS

A set of spare modules and components will recommended by DACCO.

Spares make money. Find a way to sell spares with the system, or later.

During the Design Review, personnel will define a spares philosophy to include:

1. Portions of the system with highest priority for operation
2. Components with highest expectation of failure in normal operation
3. Elements of the system with spares on module level, those with spares on the component level, and those with no spares to be provided.

The spares recommendations will take the above decisions into account, and will present a detailed and priced list for consideration.

4.10 SCHEDULE AND TASK ANALYSIS

This section presents the schedule we will follow in the development and delivery of the Data Acquisition System and the work breakdown structure (WBS) of the program. The schedule for the system is shown in Figure 4-1, and the WBS is shown in Figure 4-2.

Do not save these until last! Both should be prepared very early in the proposal preparation sequence, and modified as necessary to reflect your development of a detailed proposal.

The schedule is divided into four time phases. Phase 1, program planning, will occur in month 1, and will include release of an internal budget, developing program staffing assignments and schedule, holding an internal kickoff meeting to coordinate activities of the departments involved (engineering, software, production control, quality assurance, purchasing, contracts, etc.), and placing requisitions for equipment. This activity is shown on line ____ of the schedule. This phase is necessary to avoid inefficiencies later in the program.

Phase 2, which occurs in months 2-8, assembles the system. Line ____ of the schedule shows procurement of all vendor items, line ____ shows manufacture of all standard products produced by our normal production group; _____ shows the design and manufacture of unique hardware (brackets, I/O panels, cables, etc.); line _____ shows assembly of the equipment into the respective enclosures; lines _____ show development and test of two new circuit boards (one to _____, and the other to _____); and lines _____ show software design and development and database definition.

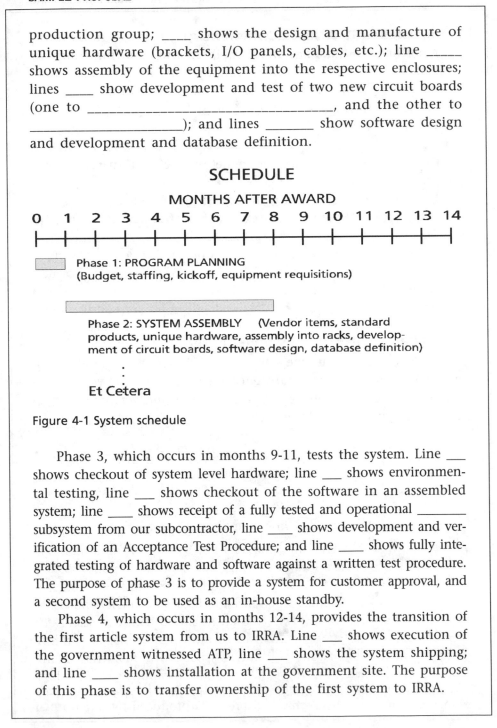

SCHEDULE

MONTHS AFTER AWARD

Phase 1: PROGRAM PLANNING
(Budget, staffing, kickoff, equipment requisitions)

Phase 2: SYSTEM ASSEMBLY   (Vendor items, standard products, unique hardware, assembly into racks, development of circuit boards, software design, database definition)

Et Cetera

Figure 4-1 System schedule

Phase 3, which occurs in months 9-11, tests the system. Line ___ shows checkout of system level hardware; line ___ shows environmental testing, line ___ shows checkout of the software in an assembled system; line ____ shows receipt of a fully tested and operational _____ subsystem from our subcontractor, line ___ shows development and verification of an Acceptance Test Procedure; and line ___ shows fully integrated testing of hardware and software against a written test procedure. The purpose of phase 3 is to provide a system for customer approval, and a second system to be used as an in-house standby.

Phase 4, which occurs in months 12-14, provides the transition of the first article system from us to IRRA. Line ___ shows execution of the government witnessed ATP, line ___ shows the system shipping; and line ____ shows installation at the government site. The purpose of this phase is to transfer ownership of the first system to IRRA.

The detail you use here is an indication to the customer that you really understand the requirement. Of course, you can't write and price the proposal without thinking these details out, so describe them clearly.

Other activities necessary to support the effort are also shown in the schedule. They include CDRL items, program reviews, and in-process reviews. CDRL numbers are listed in the TASK column on the schedule and discussed fully in section 5 of this proposal. The one schedule deviation occurs for CDRL A013, ATP preparation. It has been our experience that an Acceptance Test Procedure cannot be completed until a new system is operational. We plan to start it after the purchased equipment and technical manuals arrive, and plan to complete it during the first month of system testing.

If you have a good, logical reason to deviate from the customer's stated schedule, say so in a good, logical way. The customer is not likely to be upset, although, of course, he may not accept your deviation.

The work breakdown structure to be used in managing this program is shown in Figure 4-2. Internal cost estimating, time phasing of resources, and organization of tasks is done based on this WBS.

One item with a long lead time for the first article system has been identified: the _____ subsystem provided under subcontract. The subcontractor will be included in all design reviews held with IRRA and all management reviews held, and will provide a monthly status report that will be forwarded to IRRA. We fully expect that if a problem arises it will be identified early.

No tasks or services, other than procurement of standard equipment, subcontract of the antenna system, and use of contract environmental test facilities, are planned for subcontract in this program.

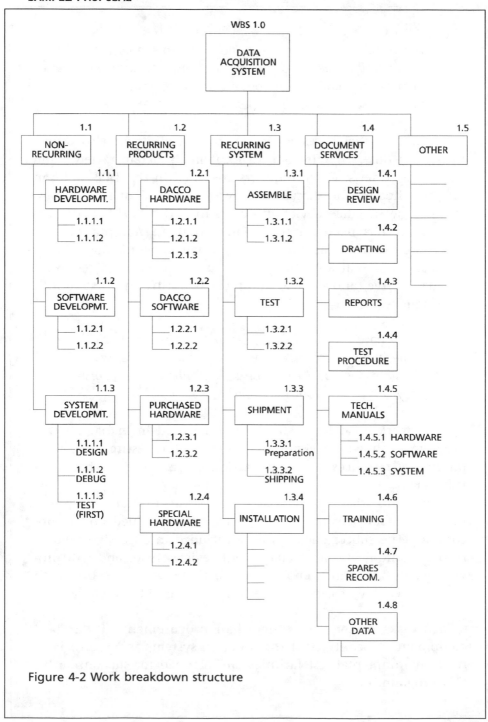

Figure 4-2 Work breakdown structure

SECTION 5: SPECIFICATION COMPLIANCE

How can we respond to an unattainable or impractical speci fication? There are several ways to do it without closing the door in the customer's mind... so let's explore a few of them.

Case 1: IRRA specified a maximum weight of 40 pounds for one of the assemblies. One of our competitors can meet this, but our unit weighs 45 pounds. We responded in such a way as to hurt the competition:

We would like to see the weight limitation raised from 40 to 45 pounds in order to allow us to propose our standard product. When the product was designed 3 years ago, we made a value judgment to increase the performance of our product at the penalty of a slightly greater weight than our original design goal of 40 pounds. The additional weight allowed us to incorporate several features which are not available in lighter weight units, namely:

a. More rugged construction to insure greater reliability in operating environments such as yours
b. Modular construction to enable you to expand the functionality of the equipment at some later date with minimum cost and delay, and
c. Additional component circuitry to lower the stress on critical components, thus increasing your mean time between failures.

We feel certain that these product improvements more than justify the weight differential, and are confident that such deviation from your specification is acceptable.

Case 2: Here again, the IRRA wrote a specification around a competitor's equipment. Our salesmen told us that it was not an expression of favoritism, but merely "insurance that someone could be compliant." Rather than make a forbidding looking list of all the items in which we deviated from the specification, we said:

211

Our response to your paragraphs ____ through ____ is based on use of our standard Model _____, which offers an improvement in performance over the equipment specified. Characteristics of our Model ____ are shown in Section _____. Use of this equipment will give you the following advantages:

a. _____
b. _____
c. _____

Case 3: IRRA asked for an impossible operating characteristic. The laws of physics prevent you (or anyone) from meeting the specification. We were placed in a very uncomfortable position here; either we look awfully dumb by taking exception when someone else can meet the specification, or look awfully smart by showing our understanding of the problem when the competition doesn't. The problem was to make certain are that we look awfully smart, and that we don't make the customer look awfully dumb! How?

We find it impossible for this specification to be met in the precise form in which it is stated, for the reasons which are shown in paragraph _____ of this proposal. Some alternatives are offered:

1. We can meet a specification of ____ if all other conditions remain as stated, or

2. We can meet this specification if the _____ is changed to _____ or if the _____ is changed to ____. We will be pleased to work with IRRA to select the most desirable performance trade-off.

Generally, you will want to list all specification paragraphs and respond to each. The exception is Case 2, above. Often a one page matrix response is the most appropriate means of assuring the Customer of your compliance.

SECTION 6: CORPORATE CAPABILITIES

While most persons who are involved with high speed data gathering, storage, and processing technology are quite familiar with

DACCO as a leader in technology, it is perhaps wise to review our origins and our present position and capabilities.

This will accomplish these purposes for the reader, and assure prospective new customers of our ability to solve difficult problems in the real-time data handling world.

6.1 BUILDING

Located on a 30-acre tract in Electronics Valley, Colorado, the Company is housed in a 100,000 square foot building of modern steel, masonry, and glass construction, completely air conditioned for personnel efficiency as well as for equipment protection.

The facilities are flexible. Building areas are modular in design, permitting optimum use of facilities no matter what the fluctuating demands of the industry may require.

Engineering occupies 22,000 square feet of the facility. This area contains the necessary facilities and equipment for the design, development, and documentation of standard products, special products, software, and systems. Also included in this area are the environmental test lab, photo lab, model shop, and system assembly areas.

Administrative functions occupy 12,000 square feet of the facility. This includes Management, Personnel, Accounting and Marketing.

The remainder of the facility is devoted to Manufacturing. This area includes facilities and equipment necessary for the fabrication, assembly, and test of products and special equipment. Metal work and printed-wiring work are performed in house. Extensive use is made of numerically controlled equipment in areas of Manufacturing.

The present level of employment is approximately 400 people.

There is no labor union at the DACCO facility, nor has there ever been a labor strike against the facility. DACCO complies with Title VII of the 1965 Civil Rights Act and Executive Order 11246, as amended.

6.2 ORGANIZATION

In dealing with DACCO, a customer sees three major areas of activity which relate to the specific product or system. They can be pictured as follows on page 214:

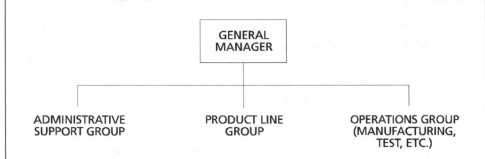

6.2.1 Product Line Group

Each product line group has all functions and activities which relate uniquely to the category of business in which they are engaged. These include:

- ☑ Design engineering, both hardware and software
- ☑ Systems engineering (as necessary), both hardware and software.
- ☑ Product line marketing, sales, technical assistance
- ☑ Product planning, et cetera.
- ☑ Program management
- ☑ Field service
- ☑ Technical support

1. DESIGN ENGINEERING

The Division has a continuing dedication to maintaining a lead in research in the areas of technology which have made them successful to date. To this end, a several-year plan for each product line group is updated, evaluated, and implemented on a year-to-year basis. Experienced engineers, programmers, and technicians work as design teams on the new products of the future.

DACCO has been more fortunate than most electronics companies in keeping valuable and experienced design personnel. Electronics Valley is a nice place to live, the Company is a nice place to work, there are no competitors in the high technology labor market nearby, and the design challenge is what engineers and technicians need. The average length of experience at this Division by design engineers is over 10 years. Software engineering has not been an integral part of DACCO business for so long, yet the emerging pattern is equally obvious.

Many companies cannot brag about the longevity of em-
ployment. If you can, do so. It tells the Customer that he is
very likely to get the benefit of experience on the program
and that it is unlikely that the team members will leave while
his program is under way.

2. Systems Engineering

For those product line groups which specialize in supplying complete turnkey systems, a systems engineering organization handles all
technical implementation of these systems from time of order until the
system is operating properly at the customer facility. This systems organization consists of engineers, programmers, technicians, assemblers,
and other support personnel.

3. Marketing

With a realization that design engineers should design rather than
be involved in sales support, yet realizing that sales engineers in remote
locations find it impossible to supply all the answers in analysis of complex requirements, the Company has created highly technical marketing
organizations to bridge the technology gap and provide effective interfaces between customers with data handling problems and those in the
Company who have solutions to those problems.

Each product line group has a sales organization. The DACCO
sales engineers are experienced (typically 8 years service); both they
and sales representatives participate in frequent training sessions.

Why mention this? Because this is the department which
has related to your customer in all those years before this
requirement come along. He knows them?

Within the plant, each product line group has an Application Engineering organization, effectively a group of staff engineers with 6
to 10 years experience. These are available to examine the application
of known techniques to solution of problems, or the development of
new techniques when nothing is available. They serve as consultants
to the Systems Engineering group and program manager on active
programs.

This is another group of your people whom your customer knows. Let the customer see their place in your organization.

4. TECHNICAL SUPPORT

Traditionally the Company has looked on product and system responsibility as a never-ending opportunity. Until the end user is trained to use equipment, armed with suitable documentation on how to use and maintain it, supplied with spares and replacement modules as needed, and assisted when the problems get too difficult or the schedule of use too demanding, the responsibility of the Company is still felt. To this end, several service functions are available.

Don't forget the people who are otherwise invisible before a contract is awarded, but who are key to successful completion of some of the important details.

A technical writing group prepares documentation on products and systems, incorporating those hardware and software definitions which are appropriate for use and repair.

A technical training organization, comprised of professionals in the business also, offers standard classes on a number of subjects relating to technology or equipment. They prepare special courses of instruction also, for presentation at our plant or at a user location.

Installation is another function which the company is eager to accept as a task. This is another way of assurance that the user is ready to handle a complex piece of hardware of software.

Along with this is field service, both hardware and software, through one of the several service locations in the United States. Various types of service contract arrangements are available, starting with on-call time and material contracts and expanding to full-time on-site service (or even service and operation).

A spares and repairs administrator in each product line group maintains computer access to all functions necessary to locate a spare part immediately, and/or to expedite the repair of a piece of equipment which is returned to the Company.

5. CONTRACT ADMINISTRATION

Another group, this one more visible to the customer, is Contract Administration. Each contract is shepherded through all administrative details by this group, and each customer is able to ascertain status and progress through the Contract Administrator assigned to the contract. Again, longevity is a key to effectiveness; the typical length of service is greater than 10 years in this group.

Again, the Customer will meet this group soon, and will deal with them for the life of the program. Introduce them!

6.2.2 Operations Group

All manufacturing and test is provided through a single group for improved control and efficiency. The Operations Group includes order entry, procurement, fabrication of metal and printed circuit elements, assembly, test engineering and production test of products, and shipment.

Separate from the Operations Group in reporting structure but key to the proper manufacture of superior equipment is the Quality Assurance group, with responsibilities all the way from engineering design review to the shipment of a finished product. This group monitors production and test, assuring proper calibration of tools and instruments, and verifying the quality of the finished product.

Important to this group is the Quality Circle technique, whereby another element of employee motivation and participation in quality is used throughout the Operations Group.

Reliability analysis of specific products, and verification of such analysis, is another aspect of total responsibility which the Company accepts.

6.2.3 Administrative Support Group

Within any company, certain administrative functions must be handled. To quantify the differences between such functions in different companies is difficult. However, at the Division those functions which the customer seldom sees are instrumental in keeping the visible functions running efficiently. At the Division, employee longevity and job satisfaction are higher because of the job done by Personnel, Accounting, Plant Maintenance, Telephone Operators, and other "invisible" organizations. In a sense, this indeed quantifies the effectiveness of those organizations.

In production of technical documentation relating to a job, a Word Processing group handles technical writing and editing through several computer-controlled terminals and printers.

6.3 ENGINEERING FACILITIES

6.3.1 Hardware and Software Design

Because correct, efficient, quick response design is the key to high technology programs of DACCO, the company has invested in unusually powerful engineering design facilities. All of these facilities are chosen to benefit the customer, as they enable us to serve program needs such as this one at IRRA properly at reasonable cost.

6.3.2 Top-Level Design

For the system level and early stages of product level design, we have 10 IBM PCs available to engineers, each equipped with _____ software. This enables the user to generate and evaluate conceptual designs before committing them to paper.

6.3.3 Module Level Design

For the bulk of detailed design effort, we have eight terminals on a Valid computer-aided engineering (CAE) system. This facility enables a designer to take a design from schematic to finished product in this sequence:

1. The designer enters a schematic of the desired module into the system.

2. The system emulates operation of a circuit based on the schematic. At this point, the validity of design is evaluated, and problems of speed, timing, et cetera are detected. The system library includes details on most logic chips used in our typical designs.

In cases where a standard chip from the library is not used, the system can interconnect a "real chip" into the emulation process.

3. Based on interaction at the emulation stage, a corrected schematic is generated by the system.

4. If a PROM is required in the design, the map of that PROM is generated; the designer can "burn" a PROM directly at this point if desired.

5. The system generates a wire list from the corrected schematic. This list is output to floppy disk.

6. A multi-wire vendor uses the floppy disk to construct a module which meets the design criteria established by this process. Response time is very short on multi-wire boards which are designed in this way.

This is important to the engineers at the Customer agency who will evaluate your proposal. It assures them of efficient design and quick production turn around.

The net result is extremely valuable to IRRA, of course. Designs are correct and fast, PROMs are a product of the system, debugging time has almost disappeared, deliveries are faster, and costs per unit of performance are lower than with any other design sequence.

6.3.4 System-Level Design Verification

In order to provide a tool for system-level design verification before the actual system is assembled, our "glass room" system is available to engineers and programmers. This is a fully operating computer system, VAX-based, with an assortment of front-end products, such that a designer or programmer can evaluate new designs or programs quickly and efficiently.

The computer system includes:

VAX 11/780 Computer
- Tri/density tape
- Two 256-megabyte disks
- Floppy disk
- Printer
- Communications multiplexer
- Several terminals/displays

The front-end hardware includes:

Decommutator
- Expander

- DAC/Discrete Unit
- Time Code Generator/Translator
- Tape Search Unit
- Multiplex Processor
- Instrumentation Tape Transport

Brag about the special capabilities which are meaningful to the Customer. Possibly they are not an integral part of this program, but they show your capability to handle product and software evaluation.

6.3.5 Data Processing Laboratory and Network

A computer central station with three VAX systems forms the central support function for all DACCO technical personnel. Each design engineer, system engineer, programmer, field service engineer and programmer, applications engineer, and otherwise-involved engineer or programmer has access through one of the 160 terminals located in our company facilities throughout the United States.

These VAX computers are linked by Ethernet to the glass room system, the Valid system, and (on an ad hoc basis) to systems under construction and test. In addition, Ethernet is linked to a large microprocessor emulation and support system. We can support development on several microprocessors and bit/slice or word/slice processors, including:

a. 68000 family
b. TMS 320 family
c. Intel family
d. Analog Devices family

The engineers at your customer agency need to know this.
It is their assurance that your engineers, their counterparts, are adequately equipped.

6.3.6 Software Support

In addition to the system unique software which comes with the above referenced equipment, we have:

1. Operating Systems
VAX/VMS Operating System
RSX Compatibility Mode
RT Operating System Emulator

2. Languages
Microtec Assembler
Modulo-2 Compiler
TMS 32010 Cross Support Assembly
C Compiler
Archimedes Software
C Cross Compiler
FORTRAN Compiler and Help Library
Intermetrics C
Cross Compiler for MC68000
TMS32010 Assembler
Prolog V1.5
Metastep Language
Pascal Compiler and Run Time Library

3. Engineering Tools
Model Simulation Tool
Gate Array Tool
Software Design and Documentation Tool
Expert System
Circuit Analysis Package
Signal Analysis Package
Quality Analysis Tool
Automatic Map and Zap Entry
Computer Aided Software Control and Development Environment

Programmers at the customer agency love this!

6.4 PRODUCTION FACILITIES

1. **Printed Wiring Boards**—DACCO has a self-contained in-house facility to manufacture printed wiring boards. Plating lines provide the copper, tin, lead, nickel and gold plating required, and a modern dark room is employed to apply, expose, and develop the dry

film photo resist. State-of-the-art computer-controlled drilling and routing machines are used to drill the plated-through holes and route the outside configuration of the printed wiring boards. A fusing machine completes the process by reflowing the tin lead to eliminate shorts between land areas.

2. Metal Fabrication—Sheet metal components, ranging from simple panels and brackets to sophisticated electronic cabinets, are built in-house using numerically controlled shears, punch presses, press brakes, and welding machines.

3. Machine Shop—Computer-controlled lathes, milling machines, precision boring, and grinding machines are employed to machine to extremely close tolerances (within 40 millionths of an inch on a production basis). All specialized tooling to support standard products, modified standard products, prototype, and custom products is designed and built by a short-run shop. This group contains most of the craftsmen within the organization. This expertise is shared with the rest of the factory to introduce new products or technology and to help solve complex fabrication/assembly problems.

4. Metal Finishing—A modern metal plating area provides the in-house capability to anodize, iridite, and cadmium plate components to customer requirements. A paint shop with modern water wash spray booths and walk-in baking ovens provides the capability of painting to federal standards. Silk screening, solder masking, and engraving capabilities are also provided.

5. Assembly—Transformer, inductors, filters, and cable harnesses are built in the magnetics and cable harness area. An encapsulation lab provides varnishing, potting, and vacuum impregnation for all assembly groups.

Microprocessor-controlled component locators are used in the assembly of printed wiring boards, and an in-line wave soldering machine and a component lead cutter complete the process. Most assemblies are tested on automatic computer controlled test equipment. Numerical controlled wire wrap machines are used to wire several thousand connections daily and a computer-controlled back plane tester is used to check all terminations.

Throughout the production area, numerically controlled equipment is used for such functions as metal fabrication, printed circuit board drilling, backplane wire wrapping, and automatic checkout. All of this makes the final product more accurate, less expensive per unit of performance, and easier to schedule.

6.5 SPECIALIZED FACILITIES

Supporting the engineering and production activities at DACCO are specialized laboratories as follows.

1. **Materials and Processes Laboratory**—Develops encapsulating and potting techniques to special engineering requirements, and evaluates physical properties of electronics packaging materials. In addition, develops chemical and metallurgical processes required by Production, performs daily analysis of chemical baths, and conducts destructive tests on manufactured samples.

2. **Calibration Laboratory**—Maintains and calibrates inspection and test instruments used by Engineering and Production in accordance with military equipment specifications. DACCO laboratory standards are traceable to the National Bureau of Standards.

3. **Reliability Analysis Laboratory**—Evaluates components to be used in DACCO products; assesses their use in circuits and with other components; analyzes equipment failures and recommends corrective action; recommends procedures for high reliability programs; conducts like testing, stress analysis, and failure mode and effect studies. Lab utilizes high powered microscopes, X-ray, and photography techniques for study and analysis of components and materials used in DACCO equipment.

Do you need all of this on your proposed program? Tell them anyway!

6.6 TEST FACILITIES

1. **Products**—Due to the unique characteristics of each product, it is not practical to list all of the test facilities in each of the 31 stations used for telemetry product test. One of the representative products, however, is the _____ Decommutator. To test this requires:

☑ Simulator
☑ Oscilloscope
☑ Special test fixture

☑ Serial Data Generator
☑ Volt ohmmeter
☑ Logic Analyzer

Each product is tested in accordance with the latest procedure; all records relating to inspection and test are retained for 4 years and are available for perusal on request.

2. **Systems**—A large amount of equipment is available to the system engineer and programmer on an as-required basis for test of each unique system. On the Data Acquisition System, we will list the required equipment on our acceptance test procedure for IRRA approval, and will verify availability and calibration status prior to start of tests.

PROPOSAL APPENDIX A

Because of the details of certain supporting literature, they are placed in the Proposal Appendix. The appendix includes sales brochures on the standard units which are proposed, plus a technical paper on the technology which is proposed.

Specifically, these references are included:

❖ _____
❖ _____
❖ _____
❖ _____
❖ _____
❖ _____

PROPOSAL APPENDIX B
SYSTEM EXPERIENCE

In a typical year, DACCO develops and delivers 20 to 25 systems of the same relative complexity as the proposed IRRA system. Some of the more recent contracts which have similarity to this are described in the Appendix. In all cases, further details are available to IRRA evaluators on request.

Contract Description: Boeing Aerospace ASAT
Customer: Boeing Aerospace
Site Installation: Edwards Air Force Base
Contract Number: P10610/F04701-80-C-0040

This system supports development and testing of the "Anti-Satellite Missile" (ASAT) during launch from the F-15 Aircraft through intercept. The hardware and software system was designed to provide decommutation of complex dynamically switchable formats, and real-time processing with graphic and alphanumeric data display. This turn key system provides complete support of real-time flight test operations including setup, data base generation, display definition, data recording, and real-time processing and display of critical data.

DACCO was responsible for the total system design, and delivered it as a turnkey system including the display software. Using a highly sophisticated man-machine interface, the menu/form technique along with help messages provide an easy-to-learn and easy-to-use software system.

Contract Description: Martin Marietta Small ICBM
Customer: Martin Marietta, Denver, Colorado
Site Installation: Denver, Colorado
Telephone No: (303) 971 6702
Contract Number: ND6-128001

Five medium size PCM Data Reduction Systems were shipped to Martin Denver in support of the Small ICBM Program. The last two systems were completed and ready to ship 2 months ahead of schedule. The first system had a short delivery requirement; it was also delivered on schedule to support the initial testing of the Small ICBM.

DACCO was responsible for the system design, and delivered a turnkey system. The system provides complete support of test operations including setup, database generation, display definition, data recording, processing, display, and printout of critical data.

Contract Description: Navy AERO Systems
Customer: Naval Surface Weapons Center, Dahlgren, VA
Site Installation: General Dynamics, Pomona, CA;
Contract Number: N6092186C-A352

Each of the four Automatic Engineering Readout (AERO) Systems is a PAM/PCM Assembly, mounted in nine portable containers for movement onto and off missile launching ships. An AERO System supports the Navy's standard missile program, and is especially tested to verify operation with missile telemeters (PAM and PCM).

The receive/record stack of equipment in AERO controls two auto-tracking antennas (not a part of the system). Telemetry carriers from these antennas or from one or two AN/SYR-1 antennas are received and demodulated in AERO. The telemetry data plus time and ship's audio are stored on an instrumentation tape. One channel is recorded also on a digital cassette recorder. This entire stack of equipment is powered via an uninterruptible power supply (UPS), to assure coverage of an entire missile flight even if ship's power is lost.

The process/display stack of equipment captures PAM or PCM data. The PAM Synchronizer or the PCM Bit Synchronizer prepares data for the Decommutator. Decommutated data enters a Display Station, a processor/display assembly built around the DEC MicroVAX II Computer and DEC color graphic display terminal. Missile data is displayed numerically in binary, decimal, octal, or engineering units; alternatively, a scrolling graphic display of one to four channels is provided. The display station operates in real time or on playback data from the computer disk.

CHAPTER

14

Sample Proposal Instructions from a Government Procurement

The following set of instructions is excerpted from a U.S. Navy procurement, and illustrates the general format and typical content of U.S. Government proposal instruction and evaluation sections. These instructions are not applicable in other procurements, of course; your customer's unique procurement package is the guiding document for a specific proposal.

1.1 PROPOSAL

1.1.1 General Proposal Requirements

Each offeror's response to the solicitation should reflect in detail the method, process, and/or other aspects or elements proposed for performance of the effort specified in this solicitation. The degree of such detail and the depth of the offeror's analysis will be important factors affecting the Government's judgment as to the offeror's comprehension of the effort to be performed. Only relevant material should be included, however. References which are cited to support analysis or test results must be included as part of the proposal. The proposal must be prepared in accordance with the instructions and format given herein and any proposal not complying with these instructions may be considered unacceptable.

The proposal to be submitted shall be prepared on standard 8 1/2" x 11" paper, single spaced with foldouts as required, with each fold of the foldout page counted as a single page and no single foldout to exceed 17" x 11". The type to be used shall not exceed fifteen (15) characters/spaces to the linear inch and shall not exceed six (6) lines/spaces to the vertical inch. There is no print size limitation on the cost volume. Each section within a volume shall start on a new page. The original and appropriate number of copies of all volumes as indicated below shall be furnished.

You are encouraged to use the least expensive form of reproduction that will provide legible copies. Elaborate binding is not required.

In presenting material in the proposal, the offeror shall follow the general rule that quality of information is significantly more important than quantity.

The proposal shall contain all pertinent information in sufficient detail to permit evaluation. No appendices, attachments, etc., will be permitted to the technical proposal unless otherwise directed below.

The proposal shall be organized into two (2) volumes, as indicated below. Volumes I and II and the completed RFP shall be submitted by the closing date of this solicitation. Each volume shall contain a table of contents, which will not be counted as part of the page limitation, and shall be dated and serially numbered. Each page and paragraph shall be numbered so that its location is traceable to its appropriate volume and section.

The two (2) volumes must be separately bound. Cost and pricing data shall appear only in volume II. Noncompliance with this requirement will result in rejection of your offer.

Volume	Number of Copies	Title
I	Original + 4	Technical Proposal
II	Original + 1	Cost Proposal

1.1.2 Specific Proposal Requirements

1.1.2.1 VOLUME I—TECHNICAL PROPOSAL

A. TECHNICAL APPROACH

The technical proposal shall contain all information necessary for a thorough evaluation and for making a sound determination as to whether or not the product proposed meets the requirements of the Navy. To this end, the technical proposal should clearly demonstrate that the prospective contractor understands and can and will comply with the requirements of the Contract Specification and the Statement of Work (SOW). Statements paraphrasing the Specification or SOW or parts thereof are considered inadequate. So are phases such as "Standard procedures will be employed" or "Well known techniques will be used" or "The offeror concurs". All of the technical factors cannot be detailed herein. However, the technical proposal must be sufficiently complete to show how the offeror's approach will comply with the applicable specification and the SOW. The proposal shall include explanations of the techniques and of the procedure to be followed in the proposed approach. Elaborate format, binders, and "prettiness" are not necessary. Clarity and completeness of discussion and analyses should be oriented towards competent technical personnel who may have no prior familiarity with the specific methods of analysis and display chosen by the offeror. Data previously submitted, if any, will not be considered and should not be relied upon or incorporated in the technical proposal by reference.

Volume I shall be organized as follows. No price/cost information shall appear in the Technical Proposal.

Section I—System Implementation

The offeror shall describe how he proposes to meet the requirements stated in the SOW/Specifications.

Section II—Experience

The offeror shall describe his experience in providing telemetry data conversion systems.

Section III—Risk Assessment

The offeror shall describe his assessment of the areas of probable risk that tend to impact the delivery schedule, reliability, and acceptability of the final product. Also, the offeror shall describe alternative actions that can be taken to reduce any adverse impact.

1.1.2.2 Volume II—Price/Cost Proposal

For the purpose of performing a cost realism analysis, the price/cost proposal shall provide the following information:

1. Direct Labor

(a) Provide a breakdown of proposed labor rates in the following format:

Labor Category	Base Rate	Base Year	Projected Rate
	(1)		(2)

(1) For each labor category, give the base rate from which the proposed rates are calculated. This rate may be either the average of the labor category as of a certain date or the average of the rates applicable to the indviduals to be assigned under this category. Provide the date on which the base rates were effective. This date should not be more than two (2) months from the date of the price/cost proposal.

(2) Provide the proposed labor rates with any applicable labor escalation. Current Navy guidance is to restrict wage escalation to 3.5% annually. If an escalation rate which is higher than 3.5% is proposed, a justification for the higher escalation rate should be provided. Labor escalation should be applied to the mid point. If a different method is proposed, provide an explanation of the methodology.

(b) Provide a breakdown of the Offeror's actual labor rate history in the following format:

	Labor Rates				
Labor Category	1997	1996	1995	1994	1993
	$	$	$	$	$

2. Labor Overhead

(a) Provide a breakdown of your labor overhead rate history and projection in the following format:

	1997	1996	1995	1994
Proposed Rate DCAA Negotiated Forward				

	1997	1996	1995	1994
Pricing Rate Actual Audited Rate				

	1993	1992	1991
Proposed Rate DCAA Negotiated Forward Pricing Rate Actual Audited Rate			

(b) Provide an explanation of any significant changes in the rate (i.e., more than one percentage point). Address the impact which this effort might have on labor overhead rates.

3. General and Administrative (G&A) Rates

(a) Provide a breakdown of your G&A rate history and your projections for the future. The following format shall be utilized.

	1997	1996	1995	1994
Proposed Rate				
DCAA Negotiated Forward Pricing Rate				
Actual Audited Rate				

	1993	1992	1991
Proposed Rate			
DCAA Negotiated Forward Pricing Rate			
Actual Audited Rate			

(b) Provide an explanation for any significant changes in the rates (i.e., more than one percentage point). Address the impact which this effort might have on G& A rates.

4. Forward Pricing Rate Agreements

Include copies, if applicable, of all approved direct and indirect forward pricing rate agreements.

1.1.2.3 REALISM OF COST OR PRICE PROPOSALS

An offeror's proposal is presumed to represent his best efforts to respond to the solicitation. Any inconsistency, whether real or apparent, between promised performance and cost or price, should be explained in the proposal. For example, if the intended use of new and innovative production techniques and their impact on cost or price should be explained; or, if a corporate policy decision has been made to absorb a portion of the estimated cost, that should be stated in the proposal. Any significant inconsistency, if unexplained, raises a fundamental issue of the offeror's understanding of the nature and scope of the work required and of his financial ability to perform the contract, may be grounds for rejection of the proposal. The burden of proof as to cost creditability rests with the offeror.

1.1.2.3.1 MINIMUM ACCEPTANCE PERIOD. Ninety (90) calendar days shall be considered the offer acceptance period unless a longer period is inserted by the offeror. However, proposals offering less than ninety (90) calendar days for acceptance by the Government from the date set for opening of offers in Item 9 of SF 33 may be considered nonresponsive.

1.2 EVALUATION FACTORS
1.2.1 General Information

Award will be made to the offeror whose proposal, based upon the Evaluation Criteria listed below, is most advantageous to the Government.

The proposals will be evaluated on the basis of general quality and responsiveness to stated requirements, the past performance of the contractor, understanding of the Statement of Work (SOW)/Specification, adequacy and soundness of the proposed technical approach for accomplishing the SOW/Specifications, and total price to the Government. Although each category will be considered individually, the final analysis of the proposals will provide for their integration, taking into account the interdependence of each with the others. Statements should be substantiated and factual. Comments such as "will meet the specifications" or "will comply with all requirements" will be considered unacceptable. Selection of offerors for further discussions and negotiations will be made of firms which are within the competitive range based on the evaluation factors listed below.

If at any stage of the negotiations all offerors in the competitive range are determined to have submitted substantially equal technical proposals, the right is reserved to notify such offerors that price will be considered the predominate factor in determining who shall receive award after submission of best and final offers.

1.2.2 General Evaluation Factors

Proposals will be evaluated and scored in accordance with the criteria set forth below. The evaluation will determine the extent to which the technical proposal is expected to accomplish the requirements set forth in the Statement of Work. All elements of the proposal will be evaluated by a team of selected personnel experienced in the area they will evaluate. The areas of "Technical" and "Price" shall be evaluated with the "Price" area weighted approximately four (4) times that of "Technical".

All technical proposals will be evaluated to ascertain whether they meet the Government's requirements as set forth in the solicitation. The three (3) considerations set forth below will be applied in evaluation of the specific criteria below.

1. Understanding the Requirements. The proposal shall indicate an appreciation of the basic requirements and their interrelationships. The proposal shall demonstrate that the SOW/Specification has been understood and has been interpreted in every respect.

2. Compliance with Requirements. The proposal shall demonstrate the extent to which the offeror has adhered to the solicitation and complied with the stated requirements.

3. Soundness of Approach. The offeror's approach and methodology shall be technically sound. The chosen approach shall be compatible with imposed schedule limitations.

1.2.3 Evaluation Criteria

Award shall be based on the following evaluation factors in descending order of importance: Price and Technical. (Price is significantly more than Technical.)

1.2.3.1 Price Proposal Evaluation Criteria

Generally, the offeror's price proposal will be evaluated for reasonableness and realism for the price of the technical approach offered, as well as to determine the offeror's practical understanding of the requirement. Price proposals will be evaluated on the basis of cost realism, during source selection, by a cost evaluation panel. This panel will receive inputs from other members of the Government evaluation team as necessary. Specific evaluation criteria are listed below in descending order of importance unless otherwise indicated. (Element (1) is approximately $1^1/_2$ times greater than Element (2).)

Element (1) Reasonableness:

The criterion of "Reasonableness" is used to evaluate the acceptability of the offeror's price proposal in relationship to (a) the other acceptable offers within the "Competitive Range" and (b) the Government price estimate for the supplies required.

Element (2) Realism:

The criteria of realism is used to evaluate the compatibility of costs with proposal scope and effort, including:

Consistency with the SOW/Specifications, RFP, all information required by these.

Realism is based upon:

a. Supportable estimates with assumptions, learning curves, equations, estimating relationships, inflation indices, cost factors, history, etc., being clearly stated, valid and suitable.

b. Sound, rational judgment used in deriving and applying methodologies, with appropriate narrative explanation and justification.

1.2.3.2 TECHNICAL PROPOSAL EVALUATION CRITERIA

Specifically, the evaluation will encompass all elements of the offeror's efforts from the standpoint of understanding the requirements, adhering to a sound and systematic design, highlighting the risk areas and recommending solutions. Specific technical evaluation criteria are listed below in descending order of importance unless otherwise indicated. (Element (1) is approximately $1^2/_3$ times greater than Element (2), and $2^1/_2$ times greater than Element (3).)

Element (1) System Implementation:

The offerer will be evaluated on how he proposes to meet the requirements stated in the SOW/Specification, how well the offerer apparently understands the problems, how well he recognizes problem areas, what solutions he proposes to use, and how well the proposed system meets the requirement.

Element (2) Experience:

The offerer will be evaluated on his experience in providing Telemetry Data Conversion Systems. How closely the offeror's experience relates to this particular task.

Element (3) Risk Assessment:

The offeror will be evaluated on his ability to assess the areas of probable risk that tend to impact the delivery schedule, reliability, and acceptability of the final product. The offeror will also be evaluated on his proposed alternative actions that can be taken to reduce any adverse impact.

Suggestions for Additional Study

T he reader who wants to delve
deeper into some of the specific areas of high-technology marketing
will find food for thought in these publications:

MARKETING HIGH TECHNOLOGY
Dr. William H. Davidow
The Free Press (Macmillan), 1986

The author shows high-technology marketing as a crusade,
with dedication to the product and commitment to the customer satisfaction as the necessary ingredients. He was in computer marketing at Hewlett-Packard, and then wrote the book based
largely on his experiences and observations as vice president of
marketing at Intel.

MARKETING MANAGEMENT
Dr. Philip Kotler
Prentice-Hall, 1988

Dr. Kotler shows marketing as a company-wide function,
where each division and work discipline has a unique contribution. It is not limited to the high-technology arena, but has
concepts and techniques which apply to this area.

MARKETING YOUR INVENTION
Thomas E. Mosley, Jr.
Upstart Publishing, 1992

The book shows how to evaluate the marketability of a product before it is even on the drawing board. It lists ten ways to commit suicide in product development and introduction.

REVOLUTIONIZING PRODUCT DEVELOPMENT
Steven C. Wheelwright and Kim B. Clark
The Free Press (Macmillan), 1992

Both authors are professors at the Harvard Business School, and they emphasize the business-related aspects of product development. The reader is led through processes which should speed up the schedule and lower the cost of product development and introduction.

INSTRUMENT CATALOG (Annual)
Hewlett-Packard, Tektronix, or another well known high-volume, high-technology manufacturer.

This will remind the reader of the techniques and words which the experts use in product introduction.

HOW TO ADVERTISE
Sandra L. Dean
Enterprise Publishing, 1980

This is an overview of advertising media and techniques which are available to a small business. It covers consumer products as well as high-technology devices.

BUSINESS PUBLICATION
ADVERTISING SOURCE
(A monthly publication)
SRDS, Wilmette, IL 60091

Lists all periodicals which advertise products in the USA, and shows the content and subscriber base of each as well as other details of interest to a potential customer.

238

BUSINESS AND SALES PRESENTATION
Malcomb Bird
Van Nostrand, 1990

Mr. Bird examines the many types of visual aids which are available in product or system presentations. He emphasizes the psychology of presentation, and shows the importance of planning a routine presentation as well as anticipating and planning ways around unexpected difficulties.

EXHIBIT MARKETING
Edward A. Chapman, Jr.
McGraw Hill, 1987

An in-depth discussion of how to display high-technology products in a trade show. The author takes a unique approach to each of four levels of participant, from the first-time exhibitor in a small booth to the large-corporation exhibitor.

THE WINNING PROPOSAL:
HOW TO WRITE IT
Herman Holtz and Terry Schmidt
McGraw Hill, 1981

This book is oriented primarily to the writers of large proposals to the U.S. Government, but it has techniques and methods of general value on other types of proposals as well. It defines the various types of government procurements, and discusses techniques for response to each. It discusses also the unsolicited proposal.

PROCEEDINGS OF A
HIGH-TECHNOLOGY CONVENTION
(Annual)

One example: Proceedings of the International Telemetry Conference Published by the Instrument Society of America, Box 12277, Research Triangle Park, NC 27709. This will show the reader a large number of examples, both good and bad, of high-technology technical presentation.

Index